T308

Environmental monitoring, modelling and control

Block 1
Potable water treatment

The Open University

prepared for the course team
by Suresh Nesaratnam

This publication forms part of an Open University course T308 *Environmental monitoring, modelling and control*. Details of this and other Open University courses can be obtained from the Student Registration and Enquiry Service, The Open University, PO Box 197, Milton Keynes MK7 6BJ, United Kingdom: tel. +44 (0)870 333 4340, e-mail general-enquiries@open.ac.uk

Alternatively, you may visit the Open University website at http://www.open.ac.uk where you can learn more about the wide range of courses and packs offered at all levels by the Open University.

To purchase a selection of Open University course materials, visit http://www.ouw.co.uk, or contact Open University Worldwide, Michael Young Building, Walton Hall, Milton Keynes MK7 6AA, United Kingdom for a brochure tel. +44 (0)1908 858785; fax +44 (0)1908 858787; e-mail ouwenq@open.ac.uk

This course has been printed on Savannah Natural Art™. At least 60% of the fibre used in the production of this paper is bagasse (the fibrous residue of sugar cane, left after the sugar has been extracted) and the balance is softwood fibre which has undergone an oxygen bleaching process.

The Open University
Walton Hall, Milton Keynes
MK7 6AA

First published 2005. Second edition 2006.

Edited and designed by The Open University.

Typeset in India by Alden Prepress Services, Chennai.

Printed and bound in the United Kingdom by Hobbs The Printers, Brunel Road, Totton, Hampshire.

ISBN 978 0 7492 1860 7

2.1

CONTENTS

Learning outcomes   5

Introduction   7

1   River quality and water treatment   8
   1.1   Introduction   8
   1.2   The General Quality Assessment scheme   8
   1.3   River classification in Scotland   14
   1.4   River classification in Northern Ireland   16
   1.5   Statutory water quality objectives   16
   1.6   River quality monitoring for pollution   18
   1.7   History and water quality   18
   1.8   The Drinking Water Inspectorate   21
   1.9   Water by-laws   23
   1.10   Water treatment   25
   1.11   Membrane processes   35
   1.12   Treatment of water from underground sources   36
   1.13   Similarities and differences between water treatment and sewage treatment   37

2   The Water Framework Directive   39
   2.1   Introduction   39
   2.2   Key factors   40
   2.3   Environmental objectives   46
   2.4   Surface water classification   47
   2.5   Groundwater classification   48
   2.6   Monitoring programmes   49
   2.7   General aspects   51

3   Pretreatment   52

4   Coagulation and flocculation   55
   4.1   Introduction   55
   4.2   Coagulants   58

5   Theory, principles and methods of clarification   63
   5.1   Introduction   63
   5.2   Settling basins – types and operation   67
   5.3   Practical considerations and choice   68
   5.4   Air flotation   68
   5.5   Selection of settling basin (clarifier) type   69

6   Filtration   71
   6.1   Introduction   71
   6.2   Rapid gravity sand filters   72
   6.3   Pressure filters   73
   6.4   Slow sand filters   73
   6.5   Other filters   74

7   Disinfection   75
   7.1   Introduction   75
   7.2   Chlorine as a disinfectant   75
   7.3   Other disinfectants   79
   7.4   Other forms of disinfectants   79
   7.5   On-site electrolytic chlorination (OSEC) systems   79

8	Industrial wastewater treatment		81
	8.1	Introduction	81
	8.2	Trade effluent control	81
	8.3	Trade effluent sampling	86
	8.4	Waste minimisation	88
	8.5	Reducing wastage of water in industry	91
	8.6	Treatment options to enable reuse	93
	8.7	A waste minimisation programme	95
	8.8	Case studies on water reuse and waste minimisation	100
9	Analysis of water		107
	9.1	Sampling	107
	9.2	Remote monitoring of water quality	109
	9.3	UK emergency procedures to protect water quality	114
10	Waterworks waste and sludges		123
	10.1	Introduction	123
	10.2	Co-disposal of waterworks sludge with sewage sludge	124
	10.3	Constituents of sludge important for agriculture	124
11	Computer modelling in the water industry		126
	11.1	Introduction	126
	11.2	River modelling	126
	11.3	Underground waters	128
	11.4	Computational fluid dynamics	129
	11.5	Water distribution system	130
	11.6	Complete water treatment model	131
	11.7	River quality modelling	131
12	Emergency water supply systems		153
	12.1	Introduction	153
	12.2	A possible treatment system	153
	12.3	Portable water purification equipment	157
	12.4	Emergency sanitation	160

Appendix 1: Characteristics of the quality of surface water intended for the abstraction of drinking water — 166

Appendix 2: EU Directive on the Quality of Water Intended for Human Consumption (98/83/EC) — 168

Appendix 3: The Water Supply (Water Quality) Regulations 2000 — 173

Appendix 4: Monitoring — 177

Appendix 5: Trade effluent discharged to the sewer: recommended guidelines for control and charging — 179

Answers to self-assessment questions — 184

References	218
Acknowledgements	219
Index	221

LEARNING OUTCOMES

The aims of this block are:

- to introduce the subject of river quality, and its regulation and monitoring in the UK
- to show the principles involved in treating underground and surface waters for potable supply
- to consider the legislation and regulations pertaining to water treatment, drinking water quality and water pollution control
- to consider minimisation of water use, and water reuse, using examples from industry
- to indicate the type of pollution monitors used at water intakes, with particular reference to water abstracted for drinking water supply
- to describe the possible actions to take in case of an emergency in relation to water treatment and supply
- to introduce the subject of trade effluent control in the UK
- to consider the disposal of sludge from water treatment works
- to provide an introduction to computer modelling in the water sector
- to introduce possible systems for emergency water supply and sanitation.

After studying this block, reviewing the associated DVD, studying the Legislation Supplement for the course, and undertaking all the assessments, you should be able to demonstrate the following learning outcomes.

Knowledge and understanding

You should be able to:

1 describe how rivers are classified in the UK, and explain the acronyms RIVPACS, ASPT, EQI and SWQOs (SAQs 2 and 3)

2 interpret the legislation and regulations aimed at maintaining the quality of the aquatic environment and public water supply (SAQs 4–7)

3 describe the main stages in the treatment of underground and surface waters for potable supply, and show how they differ from sewage treatment (SAQs 8–13, 18, 19, 54)

4 describe the main features of the Water Framework Directive, and its impact on water management in the UK (SAQs 14–17)

5 explain the circumstances under which pretreatment of raw water is required, and outline the techniques available (SAQ 19)

6 explain the theories and principles of coagulation and flocculation (SAQs 20–25)

7 explain the theory and principles of sedimentation

8 describe the various types of settling basins (SAQ 26–31)

9 explain the theory of sand filtration (SAQs 32–35)

10 explain the principles of membrane filtration

11 describe various processes and chemicals used for the disinfection of water (SAQs 36–38)

12 explain how trade effluent discharges are controlled and charged for in the UK (SAQs 39–42)

13 describe the benefits of minimising water use and reusing water (SAQs 43–47)

14 quote examples where minimisation of water use and water reuse have led to monetary and environmental benefits (SAQs 48 and 49)

15 describe the types of pollution monitors used at water intakes, with particular reference to water abstracted for drinking water supply (SAQs 50–52)

16 describe the actions that should be taken in an emergency related to water treatment and supply (SAQ 53)

17 describe the options available for the disposal of water treatment sludges (SAQs 55–56)

18 describe simple means of modelling a pollution incident in a river (SAQs 57–61)

19 describe possible systems for emergency water supply and sanitation (SAQs 62–64).

Thinking (cognitive) skills

You should have the skills to:

20 undertake design calculations related to coagulation (SAQs 21, 23–25), sedimentation (SAQs 26–28), filtration (SAQs 32 and 35), and disinfection (SAQs 37 and 38)

21 calculate trade effluent charges for given discharges (SAQs 40 and 42)

22 undertake simple computer modelling of scenarios depicting pollution incidents (SAQs 57–61).

Professional skills

You should have the skills to:

23 analyse a given situation and devise solutions for water and wastewater problems that are presented, making reasonable assumptions where required

24 present any findings in the form of a professional technical report.

Key skills

You should have the skills to:

25 analyse data

26 use scientific evidence-based methods in the solution of problems

27 communicate effectively and professionally through the tutor-marked assignments and the end of course assessment associated with the course

28 use an engineering approach to the solution of problems.

INTRODUCTION

Unlike oil, water is a renewable resource. It is also one of our most critical resources. Civilisations have, and always will, perpetuate only in areas where there is plentiful water.

This block covers water treatment for potable use. We start by considering the quality of rivers in the UK and then move on to drinking water quality and the various treatment steps needed to produce wholesome water for human consumption. The Set Book, *Basic Water Treatment*, by Binnie, Kimber and Smethurst is the principal teaching text to be used alongside this block.

Related subjects such as waste minimisation, trade effluent control, sludge treatment and disposal, water analysis and computer modelling, and emergency water supply systems, are included.

The block assumes you have access to, and are familiar with, the matter covered in the Water blocks of T210 *Environmental Control and Public Health*, or its earlier course T237.

The water supply in England and Wales is provided by 13 water-only companies, and 10 water and sewerage service companies.

In Scotland, one body (Scottish Water) is responsible for water supply, while in Northern Ireland, the Northern Ireland Water Service has this responsibility.

Water pollution control in England and Wales is the responsibility of the Environment Agency. In Scotland and Northern Ireland, the responsible bodies are the Scottish Environment Protection Agency and the Northern Ireland Environment and Heritage Service.

Water suppliers aim to provide pure, wholesome water to us. This block, together with the Set Book, covers the treatment stages involved. The block also considers ways in which the integrity of the water source is monitored.

Modelling of pollution events is introduced, as this serves as a valuable tool in ensuring that water that is abstracted is from a 'safe' site.

The Water blocks of T210/237 are: T210 Block 3 Water pollution control; T237 Units 5–6 Water quality, analysis and management; Unit 7 Water supply and sewage treatment.

STUDY NOTE

The best way to study this block is to read the chapter(s) of the Set Book indicated at the beginning of each section. Then read the section and finally answer the self-assessment question(s) provided. Answers to the SAQs can be found towards the back of the block.

---

1   RIVER QUALITY AND WATER TREATMENT

1.1   Introduction

River water quality in the UK has been progressively improving over the last 50 years, largely due to tighter anti-pollution measures (especially as regards slurries on farms) and the upgrading of existing sewage treatment works. Anecdotal evidence of the improvement in water quality of the Thames is public knowledge, but more formal measures (e.g. through the General Quality Assessment scheme) provide indisputable evidence of this.

1.2   The General Quality Assessment scheme

The General Quality Assessment (GQA) scheme is the water quality classification scheme used in England and Wales. It covers the general chemistry, biology, nutrients and aesthetics of river and canals. There are about 8000 monitoring points and over 40 000 kilometres of rivers and canals. Sites at intervals of 6 km are sampled a minimum of 12 times a year.

Stretches of rivers and canals are classified into six chemical GQA grades (Table 1) based on the concentration of dissolved oxygen, biochemical oxygen demand (BOD), and total ammonia. The overall grade assigned to a river or canal reach (section) is determined by the worst of the three grades for the individual determinands. Table 2 describes the uses and general characteristics of each grade.

Table 1   Standards for the chemical GQA

GQA grade	Dissolved oxygen (% saturation) 10-percentile	Biochemical oxygen demand (mg l^{-1}) 90-percentile	Ammonia (mg l^{-1}) 90-percentile
A Very good	80	2.5	0.25
B Good	70	4	0.6
C Fairly good	60	6	1.3
D Fair	50	8	2.5
E Poor	20	15	9.0
F Bad	$\leqslant 20$	–	–

Source: Environment and Heritage Service

SAQ 1

Define the following terms. (If you have difficulty, please reread Section 2.2 of Block 3: *Water pollution control* of T210 *Environmental Control and Public Health*.)

(a) dissolved oxygen;

(b) oxygen deficit;

(c) biochemical oxygen demand.

Table 2   The likely uses for, and characteristics of, each of the chemical GQA grades

Chemical grade	Likely uses and characteristics[a]	
A	Very good	All abstractions Very good salmonid fisheries Cyprinid fisheries Natural ecosystems
B	Good	All abstractions Salmonid fisheries Cyprinid fisheries Ecosystems at or close to natural
C	Fairly good	Potable supply after advanced treatment Other abstractions Good cyprinid fisheries Natural ecosystems, or those corresponding to good cyprinid fisheries
D	Fair	Potable supply after advanced treatment Other abstractions Fair cyprinid fisheries Impacted ecosystems
E	Poor	Low grade abstraction for industry Fish absent or sporadically present, vulnerable to pollution[b] Impoverished ecosystems[b]
F	Bad	Very polluted rivers which may cause nuisance Severely restricted ecosystems

Source: Environment and Heritage Service

[a] Provided other standards are met.

[b] Where the grade is caused by discharge of organic pollution.

Later in this section tables give the GQA grades for biological aspects (covering invertebrate life, Table 3) nutrients (relating to N and P, Table 5) and aesthetic aspects (concerned with visual appearance and odour, Table 6).

The GQA scheme is used for periodic assessment of water quality and for observing trends over space and time. Three years' data are used to assign grades to rivers. The GQA scheme has been extended to include estuaries and coastal waters.

1.2.1   Biological assessment

Biological assessment of watercourses was introduced in T210/T237 *Environmental Control and Public Health*. Here we can consider this topic further.

Water quality is an important factor influencing the status of benthic macroinvertebrates, but other characteristics of the watercourse, such as the geology, altitude, temperature, river width and depth, and the nature of the river bed, can also affect the benthic community. Consequently, it has not been easy to compare the status of macroinvertebrates in different rivers across the country owing to the difficulties of separating the influence of water pollution from other, natural, factors.

To overcome the above problem, a computerised model was developed by the former Institute of Freshwater Ecology (now the Centre for Ecology and Hydrology). Called RIVPACS (River Invertebrate Prediction and Classification System), this model allows predictions to be made of the type of community that would be expected according to a range of natural features, assuming the river is not affected by pollution.

The model accesses a database holding biological and physical data from over 8000 monitoring sites in England and Wales, covering a wide range of environmental variables. Sampling at the sites was repeated over different seasons, and the Biological Monitoring Working Party (BMWP) score of each site was computed. A measure of the sensitivity to pollution was also obtained by determining the 'average score per taxon' (ASPT) which is the BMWP score divided by the number of taxa present. In the UK, there are over 4000 different species of aquatic invertebrates, which are grouped into classes and families. These groupings are called taxa. The BMWP system is made up of 83 different taxa, each taxon scoring between one and ten, depending on how sensitive it is to pollution. A high number of taxa is itself a good indication that the water quality is good. A BMWP score greater than 100 and an ASPT greater than 4 generally indicates good water quality. Two samples are taken – one in spring (March–May) and the other in autumn (September–November) – to take account of seasonal variation. Taxa found in the spring sample are combined with any additional taxa found in the autumn.

Together with invertebrate sampling, physico-chemical characteristics of the sites are also estimated or measured. These include distance from the source, channel width and depth, gradient of the river, altitude, velocity of flow, alkalinity, temperature, and the nature of the river bed.

On the basis of the physico-chemical variables, the RIVPACS model can predict the probability of occurrence of individual taxa, and the BWMP and ASPT values.

It is possible to compare this prediction with the actual community status to provide a measure of the extent to which the river is affected by pollution. The ratio of the observed community status to that predicted can be obtained. This is called the 'ecological quality index' (EQI).

Using the BMWP score, for example, the EQI can be calculated for a given situation, as follows:

$$EQI_{BMWP} = \frac{\text{BMWP score observed by monitoring}}{\text{BMWP score predicted by RIVPACS}}$$

EQIs can be calculated for both the number of taxa and the ASPT.

$$EQI_{\text{no. of taxa}} = \frac{\text{observed number of taxa}}{\text{predicted number of taxa}}$$

$$EQI_{ASPT} = \frac{\text{observed ASPT}}{\text{predicted by ASPT}}$$

A value for the EQI of approximately 1 indicates that the biological communities found in the river are those that would be expected under conditions of natural water quality. Lower values indicate that the biota may be stressed by pollution, drought or other causes.

The EQI for taxa and ASPT are used in conjunction with Table 3 to assign a grading for the river. If the grades differ for the two indices, the lower grade is selected. Table 4 gives a description of each grade.

Table 3   GQA biological grades

Grade	EQI for ASPT	EQI for number of taxa	Environmental quality
A	1.00 or above	0.85 or above	Very good
B	0.90–0.99	0.70–0.84	Good
C	0.77–0.89	0.55–0.69	Fairly good
D	0.65–0.76	0.45–0.54	Fair
E	0.50–0.64	0.30–0.44	Poor
F	less than 0.50	less than 0.30	Bad

Source: Environment and Heritage Service

Table 4   Description of the biological grades in the GQA scheme

Grade A – Very good

The biology is similar to (or better than) that expected for an average, unpolluted river of this size, type and location. There is a high diversity of families (taxa), usually with several species in each. It is rare to find a dominance of any one family.

Grade B – Good

The biology shows minor differences from Grade A and falls a little short of that expected for an unpolluted river of this size, type and location. There may be a small reduction in the number of families that are sensitive to pollution, and a moderate increase in the number of individuals in the families that tolerate pollution (like worms and midges). This may indicate the first sign of organic pollution.

Grade C – Fairly good

The biology is worse than that expected for an unpolluted river of this size, type and location. Many of the sensitive families are absent or the number of individuals is reduced, and in many cases there is a marked rise in the numbers of individuals in the families that tolerate pollution.

Grade D – Fair

The biology shows considerable differences from that expected for an unpolluted river of this size, type and location. Sensitive families are scarce and contain only small numbers of individuals. There may be a range of those families that tolerate pollution and some of these may have high numbers of individuals.

Grade E – Poor

The biology is restricted to animals that tolerate pollution with some families dominant in terms of number of individuals. Sensitive families will be rare or absent.

Grade F – Bad

The biology is limited to a small number of very tolerant families, often only worms, midge larvae, leeches and the water hog-louse. These may be present in very high numbers but even these may be missing if the pollution is toxic. In the very worst case there may be no life present in the river.

Source: Environment and Heritage Service

1.2.2   Nutrient assessment

Nitrate and phosphate concentrations are measured, and the results expressed as mg P l^{-1} and mg NO_3 l^{-1}. A grade from 1 to 6 is allocated for both phosphate and nitrate (Table 5). These are not combined into a single nutrients grade. 'High' concentrations are indicative of possible existing or future problems of eutrophication.

Table 5   Grade allocations for phosphate and nitrate

Classification for phosphate (Grade)	Grade limit average (mg P l^{-1})	Description
1	$\leqslant 0.02$	Very low
2	$\geqslant 0.02$ to 0.06	Low
3	$\geqslant 0.06$ to 0.1	Moderate
4	$\geqslant 0.1$ to 0.2	High
5	$\geqslant 0.2$ to 1.0	Very high
6	$\geqslant 1.0$	Excessively high

Classification for nitrate (Grade)	Grade limit average (mg NO_3 l^{-1})	Description
1	$\leqslant 5$	Very low
2	$\geqslant 5$ to 10	Low
3	$\geqslant 10$ to 20	Moderately low
4	$\geqslant 20$ to 30	Moderate
5	$\geqslant 30$ to 40	High
6	$\geqslant 40$	Very high

Source: The Environment Agency 2004

1.2.3   Aesthetic aspects

The aesthetic quality of rivers and canals is based on the following:

(a) *Litter* (e.g. gross litter, general litter, sewage and dog faeces): a standard sampling unit (an area extending 50 m along the river or canal, and up to 5 m from the water's edge, plus the river/canal and its bed) is assessed and the number of litter items of each type counted.

(b) *Oil, surface scum, foam, sewage fungus, ochre*: classified on the 'percentage cover' of the water.

(c) *Colour*: classified according to its hue and intensity.

(d) *Odour*: qualitatively assessed from the bank, and classified according to type and intensity.

The general rule is that a standard site consists of both riverbanks and the water. This is referred to as a 'two-bank site'. There will be sites where only one bank has public access, or the river is so wide that each bank should be treated as a separate site. Such sites are referred to as 'one-bank sites'. The classification of a site for aesthetic quality is one of four grades.

1   Good

2   Fair

3   Poor

4   Bad

Table 6 shows how to assign a class for each parameter surveyed for a two-bank site. The overall grade for a site is derived from the score allocated to each parameter (Table 7). The number of points allocated for each parameter is based on the relative importance given to each parameter from a public perception study. The final total score is simply the sum of the points for all the parameters.

Table 6   Classifying a two-bank site for aesthetic aspects

Litter (number of items):	Type	Class 1	Class 2	Class 3	Class 4
	Gross	0	1–2	3–9	10+
	General	0–5	6–49	50–99	100+
	Sewage	0	1–5	6–24	25+
	Faeces	0	1–5	6–24	25+

Other aesthetic parameters (percentage cover of oil, scum, foam, sewage fungus, ochre):	Class 1	Class 2	Class 3	Class 4
	0	⩾0–5%	⩾5–25%	⩾25%

Colour	Blue/Green	Red/Orange	Brown/Yellow/Straw
Colourless	Class 1	Class 1	Class 1
Very pale	Class 1	Class 2	Class 1
Pale	Class 3	Class 3	Class 2
Dark	Class 4	Class 4	Class 3

Odour	Group 1[a]	Group 2[b]	Group 3[c]
No smell	Class 1	Class 1	Class 1
Faint smell	Class 1	Class 2	Class 3
Obvious smell	Class 2	Class 3	Class 4
Strong smell	Class 3	Class 4	Class 4

Source: The Environment Agency 2004

[a] Tolerated and less indicative of poor water quality (musty, earthy, woody).

[b] Rated as indicators of poor water quality (farmy, disinfectant, gas, chlorine).

[c] Rated as indicators of very poor water quality (sewage, polish/cleaning fluid, ammonia, oily smell, bad egg (sulphide)).

Table 7   Points allocation used for aesthetic classification

	Class 1 score	Class 2 score	Class 3 score	Class 4 score
Sewage litter	0	4	8	13
Odour	0	4	8	12
Oil	0	2	4	8
Foam	0	2	4	8
Colour	0	2	4	8
Sewage fungus	0	2	4	8
Faeces	0	2	4	6
Scum	0	1	3	5
Gross litter	0	1	2	3
General litter	0	1	2	3
Ochre	0	0	0	1

Source: The Environment Agency 2004

The points allocated for each parameter are then summed to give the 'total score'. The overall grade of the site is then allocated, as follows:

Grade	Aesthetic quality	Total score
1	Good	1, 2, 3
2	Fair	4, 5, 6, 7, 8
3	Poor	9, 10, 11, 12
4	Bad	$\geqslant 13$

1.3   River classification in Scotland

In Scotland, the rivers, lochs and canals are classified according to the parameters shown in Table 8.

Notes relating to Table 8

a   Based on three years' data, minimum of 12 samples, unless there has been a significant change in circumstances (e.g. a discharge eliminated or an identified major pollution incident in a previous year) which justifies an assessment based on a lesser data set collected after a step change. In such circumstances a minimum monitoring period of 12 months must have lapsed since the change, and where there are fewer than 12 samples the significance of the step change should be confirmed by a statistical test.

Estimation of percentiles to be by parametric method, assuming DO and pH are normal distributions and BOD and ammoniacal nitrogen are log normal.

For pH the 5, 10 and 95 percentiles must be determined from the three years' data and compared with the class determining limits in the classification table. Again, the parametric percentile estimation must be made, using the method of moments, and an assumed normal distribution. [For the basis of sampling strategy, see *Handbook on the Design and Interpretation of Monitoring Programmes*, WRc report NS29, by J.C. Ellis, 1989.]

b   RIVPACS assessment based on data for one year, preferably three samples (spring, summer and autumn), minimum of two (spring and autumn).

c   Based on one year's monitoring data, preferably three samples, minimum of two. The overall class to be determined from the mean field score and mean ASPT of the individual samples.

d   Aesthetic conditions to be based on one year's data from a minimum of three observations and will be assessed and recorded during biological and/ or chemical sampling visits to programmed sampling points. Aesthetic contamination to be assessed as either discharge related (List A) or general (List B).

List A contaminants	List B contaminants
Sewage-derived litter and solids	*General non-sewage-derived litter*
Faeces, toilet paper	Builders' waste
Contraceptives	Gross litter, including:
Sanitary towels, tampons, cotton buds	shopping trolleys
	furniture
Oils	motor vehicles
Non-natural foam, scum or colour	road cones
Sewage fungus	bicycles/prams
Sewage or oily smells	

Table 8 Scottish River Classification Scheme

CLASS and DESCRIPTION	WATER CHEMISTRY[a]					BIOLOGY				NUTRIENTS	AESTHETIC CONDITION[d] (contaminants)	TOXIC SUBSTANCES	COMMENTS
	DO (% sat.) 10%ile	BOD (mg l^{-1}) 90%ile	NH_4-N (mg l^{-1}) 90%ile	Fe (mg l^{-1}) Mean	pH %ile	Lab analysed[b]			Bankside[c]	[a] Soluble reactive phosphorus (µg l^{-1}) Mean			
						EQI_{ASPT}	EQI_{Taxa}	ASPT	Field score				
A1 Excellent	≥80	≤2.5	≤0.25	≤1	5%ile ≥6 95%ile ≤9	≥1.0	≥0.85	≥6.0	≥85	≤20	No A Minor B[e]	Complies with Dangerous Substances EQSs	Sustainable salmonid fish population; natural ecosystem
A2 Good	≥70	≤4	≤0.6	≤1	10%ile ≥5.2	≥0.9	≥0.70	≥5.0	≥70	≤100	Trace/ occasional A or B[f]	Complies with Dangerous Substances EQSs	Sustainable salmonid fish population; ecosystem may be modified by human activity
B Fair	≥60	≤6	≤1.3	≤2	10%ile < 5.2	≥0.77	≥0.55	≥4.2	≥50	≥100		Complies with Dangerous Substances EQSs	Sustainable coarse fish population; salmonids may be present; impacted ecosystem
C Poor	≥20	≤15	≤9.0	> 2		≥0.50	≥0.30	≥3.0	≥15		Gross A or B[g]	> EQS for dangerous substance	Fish sporadically present; impoverished ecosystem
D Seriously polluted	< 20	> 15	> 9.0			< 0.50	< 0.30	< 3.0	< 15			> 10 × EQS for dangerous substance	Cause of nuisance; fauna absent or seriously restricted

Source: Scottish Environment Protection Agency

e   No List A contaminants, possibly minor List B litter present.

f   Traces of List A and/or occasional List B contamination, especially at easy access points.

g   List A contamination widespread and/or occasional conspicuous quantities, and/or widespread or gross amounts of List B contamination. Likely to be the cause of justified public complaints.

The annual aesthetics classification is derived from the individual spot samples in the following way. Spot classifications are assigned a numerical value:

Class	A1	A2	C
Value	1	2	4

The arithmetic mean value of the spot classes for the year is calculated and the annual class assigned using the following bands (a minimum of three spot values is required for an annual class to be assigned):

Mean value	$\geqslant 3.0$	$\geqslant 1.5$	$\leqslant 1.5$
Class	C	A2	A1

1.4   River classification in Northern Ireland

A modified RIVPACS model designed for use in Northern Ireland (taking into account that certain species seen in England and Wales are absent in Northern Ireland) is used to derive EQIs for taxa and ASPT. The biological grading system used by the Environment Agency in England and Wales is then applied. A chemical grading system, also based on the Environment Agency system, is used.

SAQ 2

Sampling at a river shows the BMWP score to be 96, with 6 taxa being present. If the predicted number of taxa and ASPT are 8 and 20, respectively, grade the quality of the river.

1.5   Statutory water quality objectives

With the Water Resources Act 1991, a system of 'statutory water quality objectives' (SWQOs) has been brought into effect in England and Wales. These define use-related standards for setting quality targets to protect specific river uses, according to local and downstream needs.

The SWQOs are identified in terms of one or more use classes which prescribe the quality of water necessary for the protection of the different uses to which the water may be put. Five use classes exist:

■   River Ecosystem

■   Special Ecosystem

■   Abstraction for Potable Supply

■   Agricultural and Industrial Abstraction

■   Watersports.

The River Ecosystem classification comprises five hierarchical classes (RE1–RE5) with the criteria as set out in Table 9 below. Statistical methods for assessing compliance with the standards have been defined in the procedure.

Table 9 River Ecosystem classification

Class	Dissolved oxygen (% saturation) 10%ile	BOD (ATU) (mg l^{-1}) 90%ile	Total ammonia (mg N l^{-1}) 90%ile	Un-ionised ammonia (mg N l^{-1}) 95%ile	pH lower limit as 5%ile upper limit as 95%ile	Hardness (mg l^{-1} CaCO$_3$)	Dissolved copper (μg l^{-1}) 95%ile	Total zinc (μg l^{-1}) 95%ile
RE1	80	2.5	0.25	0.021	6.0–9.0	≤ 10	5	30
						> 10 and ≤ 50	22	200
						> 50 and ≤ 100	40	300
						> 100	112	500
RE2	70	4.0	0.6	0.021	6.0–9.0	≤ 10	5	30
						> 10 and ≤ 50	22	200
						> 50 and ≤ 100	40	300
						> 100	112	500
RE3	60	6.0	1.3	0.021	6.0–9.0	≤ 10	5	300
						> 10 and ≤ 50	22	700
						> 50 and ≤ 100	40	1000
						> 100	112	2000
RE4	50	8.0	2.5	–	6.0–9.0	≤ 10	5	300
						> 10 and ≤ 50	22	700
						> 50 and ≤ 100	40	1000
						> 100	112	2000
RE5	20	15.0	9.0	–	–	–	–	–

Source: Department for Environment, Food & Rural Affairs

SAQ 3

What do you understand by the following acronyms, all related to river quality monitoring?

(a) RIVPACS

(b) ASPT

(c) EQI

(d) SWQOs

1.6   River quality monitoring for pollution

In order to monitor for pollution in rivers, automatic sampling and measuring systems have been developed.

There are systems for river quality monitoring which are equipped with probes for measurement of dissolved oxygen, ammonium ions, conductivity, temperature, pH, flow and turbidity. These are housed in compact, automated units and continuously monitor and record water quality *in situ*. They are able to calibrate themselves automatically and only require visiting for servicing and replenishment of solutions. The Environment Agency uses two such systems. Sherlock^{TM} is used for shallow rivers (being placed on the river bank) while Merlin^{TM} is anchored and designed to float and is used in deep rivers and lakes. Whenever the value for any of the parameters indicates a pollution incident, a sample is automatically taken and a signal is transmitted by telemetry to the duty pollution inspector who can in turn link up with the unit using notebook computers equipped with cellular-radio modem links and see what the monitors are reading. The pollution inspector can then take the necessary action immediately, e.g. call out the person in charge of the premises concerned (or suspected), call up emergency teams, etc. The Environment Agency has used Sherlock and Merlin in outlets from sewage treatment works, industrial effluent treatment plants, and paper mills.

To monitor the actual discharge of effluents, a variant called Cyclops^{TM} (because of its single observation lens!) has been developed. The major difference between Cyclops and the other two systems is that Cyclops takes a bulk sample of three litres when a pollution incident is detected.

1.7   History and water quality

READ
Set Book Chapter 1: Introduction, and Chapter 2: Quality of water

Chapter 1 of the Set Book outlines a short history of water supply and treatment in the UK. Before 1950, only minimal treatment was required, and simple gravity-flow distribution was the norm. The water had to be wholesome, with standards only for microbiological parameters. Wholesomeness was described by Windle Taylor in *The Examination of Water and Water Supplies* (a standard textbook on water quality first published in 1904):

> Water taken from a properly protected source, submitted to an adequate system of purification, can be designated as pure and wholesome if it is free from visible suspended matter, colour, odour and taste, from all objectionable bacteria

indicative of the presence of disease-producing organisms, and contains no dissolved matter of mineral or organic origin which in quality or quantity would render it dangerous to health, and will not dissolve substances injurious to health.

Chlorination was universally introduced in 1936, after a typhoid outbreak in Croydon. There had been localised chlorination at a number of sites in England, following the first use of chlorination in Maidstone, Kent, in 1897 when calcium hypochlorite was used after an outbreak of enteric fever. Chlorination of London's water supply started in 1910. Water demand increased steadily in the first half of the twentieth century, and this was catered for by increased spending on infrastructure. Water supplies were largely unmetered.

Massive changes took place over the period 1975–2000. The most significant amongst these were that the water industry was privatised, numerical standards for water quality emanating from EU directives were adopted, and metering came into being in a big way. The public also became more environmentally conscious, making the building of reservoirs by water companies very difficult.

Different approaches to meeting water demand had to be sought. More attention was paid to minimising leakage, and efficient use of water became a priority.

In Chapter 1, comparison is made with water supply in developing countries where it may be preferable to supply a large number of people with lower quality but acceptable water, than to spend limited financial resources to supply high-quality water to a small population.

Chapter 2 looks at various pollutants that may be present in water. You will have come across these in T210/T237 *Environmental Control and Public Health*. Mention is made of the Water Framework Directive, details of which can be found in the next chapter of this block.

Chapter 2 discusses the setting of drinking water quality by the World Health Organization (WHO), the EU and USA. The historical development of the WHO and EU standards are outlined. The WHO *Guidelines* for *Drinking-Water Quality* are not mandatory but are intended to be used as a basis for the development of national standards that will ensure the safety of drinking water supplies. National standards should be established after taking into consideration local environmental, social, economic and cultural conditions. At the time of writing this block (late 2004), the third edition of this publication was launched.

Chapter 2 also briefly describes the analytical methods used for the main parameters of interest in drinking water quality.

The Langelier Saturation Index as a means of gauging corrosiveness is introduced. This measures the potential for precipitation or dissolution of calcium carbonate. A positive index means that calcium carbonate will be precipitated, thus protecting the pipework from corrosion.

The standards required for potable water and the ways of ensuring that these standards can be met require a little more elaboration. Although for some time most countries have had their own standards of acceptable water quality, there has been a change towards internationally accepted standards such as those to be found in the EU. The EU standards for different compounds tend to be more stringent than the 'guideline values' proposed by the World Health Organization.

In the EU, one approach to improving the quality of drinking water has been to control the discharges to the aquatic environment, with both surface and groundwaters being covered. The EU has several directives covering the quality of water, and one of these, the Directive on Pollution Caused by Certain

Dangerous Substances Discharged into the Aquatic Environment (76/464/EEC) often referred to as ENV 131, is discussed in the T308 Legislation Supplement.

This directive indicates those substances which are to be eliminated as well as those whose discharges are to be considerably reduced. It covers discharges to surface waters while groundwater is protected by a sister directive: The Protection of Groundwater against Pollution Caused by Certain Dangerous Substances (80/68/EEC).

The next step after controlling discharges is to *set the standards* required for drinking water, and to indicate the *treatment* which will be *required*. The directive we must now consider is the Directive Concerning the Quality Required of Surface Water Intended for the Abstraction of Drinking Water (75/440/EEC; see Appendix 1 to this block).

This directive proposed standards for surface water to be used for public supply. Three categories of surface water are defined in the directive, designated as A1, A2 or A3 (Figure 1), the treatment required depending on the quality of the water being abstracted.

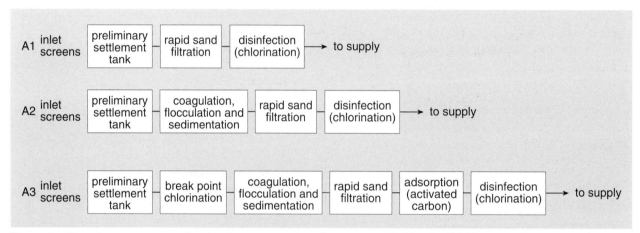

Figure 1   Constituent processes for the three categories of water specified in the EU Directive Concerning the Quality Required of Surface Water Intended for the Abstraction of Drinking Water. Note: pH correction or stabilisation is not a requirement of this directive

The directive defines A1 water as that requiring 'simple physical treatment and disinfection, e.g. rapid sand filtration and disinfection'; it defines A2 water as requiring 'normal physical treatment, chemical treatment and disinfection, e.g. prechlorination, coagulation, flocculation, decantation, filtration, disinfection (final chlorination)', and A3 water as requiring 'intensive physical and chemical treatment, extended treatment and disinfection, e.g. chlorination to breakpoint, coagulation, flocculation, decantation, filtration, adsorption, (activated carbon), and disinfection (ozone, final chlorination)'. Appendix 1 gives a list of parameters to be determined and attaches values which are set out under the appropriate treatment which the water will require. The directive uses the concept of I and G values. Those parameters having a known adverse effect on human health are given Imperative (I) values while those thought to be less adverse are given Guide (G) values. For I values, a 95% compliance is required, and for the 5% not complying, none of the samples should contain substances in excess of the I values by more than 50%. However, a relaxation can be allowed if the high values are the result of floods, abnormal weather conditions, or natural disasters.

The treatment required, as listed for each category of water quality, does not specify pH correction to stabilise the quality of the treated water to minimise corrosivity.

The above directive was incorporated into law for England and Wales by the Surface Waters (Abstraction for Drinking Water) (Classification) Regulations 1996 (SI 1996 No. 3001).

The directive will be repealed 10 years after adoption of the Water Framework Directive.

The last step in the sequence, the quality of the water supplied to the public, is covered by the Directive on the Quality of Water Intended for Human Consumption (98/83/EEC). This is a revision of the original Drinking Water Directive adopted in 1980 which came into force initially in 1985. The details of the requirements in this new directive are shown in Appendix 2.

In England and Wales, the Water Supply (Water Quality) Regulations 2000 (SI 2000 No. 3184) incorporate the above into law. Similar legislation has been passed in Scotland. In Northern Ireland, although the Water Supply (Water Quality) Regulations do not apply, British practice is adopted.

The Water Supply (Water Quality) Regulations 2000 (see Appendix 3) largely specify the limits as for the 1998 Drinking Water Directive but differ with regard to the five parameters shown in Table 10.

Table 10   Differences between the Regulations for England and Wales and the EU Drinking Water Directive

	England and Wales Water Supply (Water Quality) Regulations 2000	EU Drinking Water Directive (1998)
Colour (mg l^{-1}: Pt/Co)	20	Acceptable to customers and no abnormal change
Turbidity (NTU)	4 (at consumer's tap)	
Taste (dilution)	3 at 25 °C	
Odour (dilution)	3 at 25 °C	
pH	6.5–10	6.5–9.5

The Regulations also specify a maximum limit of one oocyst of *Cryptosporidium parvum* per 10 litres of water, and 3 µg l^{-1} for tetrachloromethane.

1.8   The Drinking Water Inspectorate

The Drinking Water Inspectorate (DWI) was formed in January 1990 with the aim of checking that companies supplying water in England and Wales were complying with their legal obligations. The main tasks of the DWI are to:

(a) Undertake technical audits of water companies. This is done to ensure that the companies are complying with statutory obligations and following good practice. There are three parts to this audit. In the first part, an annual assessment is made based on the information provided by the companies on aspects such as the quality of the water in each supply zone, water treatment works and service reservoirs, compliance with sampling requirements, and the progress made on improvement programmes. In the second part of the assessment, the individual companies are inspected with regard to the above, and an assessment of the quality and accuracy of the information is made. The third part involves interim checks which are made on the basis of information provided by the companies.

(b) Instigate action as necessary to secure or facilitate compliance with legal requirements.

(c) Investigate incidents which adversely affect water quality.

(d) Advise the Secretary of State in the prosecution of companies which have supplied water suspected to be unfit for human consumption.

Water companies are required to inform the DWI of situations that have, or are likely to have an effect on drinking water quality, or the sufficiency of supplies where such deficiencies may be a risk to the public health of consumers. Water companies are also encouraged to report situations that may lead to public concern. All notifications have to be given within specified timescales – initial notification immediately, followed by written confirmation within three days, with a final report being required within one month, depending on the severity of the situation. Such reports assist the industry to learn lessons for the future but also lead to action being taken by the DWI against offending water companies.

The DWI assesses each three-day report to determine:

- the cause (was it avoidable? what was its impact?)
- how the water company responded and managed the situation until normal conditions were resumed
- if regulations were breached
- if the water company supplied 'unwholesome or unfit' water for domestic use, such as for drinks and food preparation.

There are four basic actions taken by the DWI.

1   Where no breaches of regulations were noted and the water company acted appropriately, a letter is sent by the DWI to relevant official organisations such as local and health authorities, OFWAT, and the Welsh assembly, recording this observation and conclusion.

2   Where there were breaches of regulations and/or deficiencies in the water company's response, the DWI will provide recommendations for action by the water company. These must be carried out. Relevant official organisations are also informed.

3   Where significant or repeated breaches of enforceable regulations are noted, formal enforcement action is initiated against the water company. This is a legal process to ensure that the company takes all necessary action to prevent future breaches. Relevant official organisations are informed.

4   If the supply is regarded as unfit for human consumption due to the water company's deficient actions or response, prosecution proceedings are initiated. This can lead to court appearances, or the issue and acceptance by the company of a formal caution for committing a criminal offence. Again, relevant official organisations are informed.

(e) Provide technical and scientific advice on drinking water issues, to ministers and officials of the Department for Environment, Food and Rural Affairs, and the National Assembly of Wales.

(f) Assess and respond to consumer complaints when local procedures have been exhausted. Customers take complaints to the local WaterVoice Committee and these are taken up with the water companies concerned. If no solution is found, the complaint may be referred to the DWI, which will then request the water company to investigate and report back to it. The DWI may also require the local Environmental Health Department to investigate and report back.

(g) Identify and assess new issues or hazards relating to drinking water quality and initiate research as required.

(h) Assess chemicals and materials used in connection with water supplies. The DWI operates a scheme which assesses chemicals used in treating drinking water, and materials used in water treatment plants and distribution systems. The scheme ensures that all chemicals which are added to water are safe, and that any chemicals which leak out from construction materials are also harmless. The chemicals are also checked to see that they do not encourage microbial growth in distribution systems.

(i) Provide authoritative guidance on analytical methods used in the monitoring of drinking water quality.

In Scotland, the monitoring of the water supply is undertaken by the Water Services Unit of the Scottish Office Department of Agriculture and Fisheries. In Northern Ireland, a Drinking Water Inspector carries out this function.

1.9   Water by-laws

READ	
Set Book Chapter 14: Water demand and use	

Water by-laws were the legal means by which waste, undue consumption, misuse, or contamination of water on users' premises was prevented. These have now been incorporated into water regulations, and take effect through exerting control over the type, design and layout of fixtures and fittings. In England and Wales, the regulations are enforced by the individual water companies, and in Scotland they are applied by Scottish Water. In Northern Ireland they are applied by Water Services.

The regulations control plumbing installations, but drainage and wastewater disposal installations within the curtilage of a property are controlled by building regulations enforced by the local authority.

Chapter 14 of the Set Book looks at water demand and use. It considers the reasons why management of water demand has become critical. It also points out the role that legislation, such as the Environment Act 1995, the Water Industry Act 1999 and the Water Framework Directive, play in helping to reduce water demand.

Leakage is discussed. OFWAT publishes leakage data from water companies in terms of litres per property per day and in cubic metres per kilometre per day. This gives more useful information than simply a percentage figure.

Various ways of reducing water demand are presented. You would have met these in T210. Water companies have to assess annually their available water resources. They also have to predict water demand over a 25-year period.

SAQ 4

(a) A turbid surface water is found to contain 15 mg l^{-1} of suspended solids. Will this water have a turbidity of 15 NTUs (nephelometric turbidity units)?

(b) Would the water in (a) be acceptable for abstraction for drinking water supply?

SAQ 5

The Environment Agency decides that the quality of the water in a river is to be improved so that the river water may be abstracted for drinking water supply. It also decides that the river should be able to support game fishing. Given the information in Table 11, what maximum concentration of arsenic and iron will be acceptable in the river water? The river water has a hardness of 150 mg l^{-1} (as $CaCO_3$).

Table 11   Standards for trace substances ($\mu g l^{-1}$) for the protection of salmonid freshwater fish

Protection of salmonid freshwater fish at hardness / (mg $1^{-1}CaCO_3$)

	<50	50–100	100–150	150–200	200–250	>250
arsenic	50	50	50	50	50	50
cadmium	5	5	5	5	5	5
chromium	5	10	20	20	50	50
copper	1[a]	6[a]	10[a]	10[a]	10[a]	28[a]
	5[b]	22[b]	40[b]	40[b]	40[b]	112[b]
lead	4	10	10	20	20	20
mercury	1	1	1	1	1	1
nickel	50	100	150	150	200	200
zinc	10[a]	50[a]	75[a]	75[a]	75[a]	125[a]
	30[c]	200[c]	300[c]	300[c]	300[c]	500[c]

Source: Gardiner, J. and Mance, G. (1984).

[a] Also applies to waters which have not been specifically designated for the directive, i.e. estuaries of rivers where greater dilution can occur. It is an average value.

[b] Applies as in [a] but as a 95-percentile value.

[c] Applies to waters specifically designated for salmonid freshwater fish where the effluent enters the river. It is a mandatory value.

The following questions relate to the legislation which has been discussed.

SAQ 6

(a) If a potable water was found to have a nitrate concentration of 45 mg l^{-1} as N, would this water meet the standards set by the EU Directive on the Quality of Water Intended for Human Consumption?

(b) If a water sample from a river gave the following results, what category of treatment would you expect the water to receive?

pH = 8.5

conductivity = 1200 μS cm^{-1}

copper = 0.20 mg l^{-1}

zinc = 3.0 mg l^{-1}

phenols = 0.007 mg l^{-1}

COD = 15 mg l^{-1}

BOD = 4.0 mg l^{-1}

nitrate $NO_3^- = 35$ mg l^{-1}

SAQ 7

A water supply to consumers gave the following chemical analysis:

pH = 6.0

colour (Pt/Co scale) = 25

turbidity (NTU) = 6

chloride (as Cl) = 400 mg l^{-1}

iron (as Fe) = 0.3 mg l^{-1}

copper (as Cu) = 2.0 mg l^{-1}

calcium (as Ca) = 20 mg l^{-1}

lead (as Pb) = 75 μg l^{-1}

arsenic (as As) = 50 μg l^{-1}

Would this water be acceptable using the standards for drinking water in England and Wales?

1.10   Water treatment

READ
Set Book:
Chapter 3: Overview of water treatment
Chapter 11: Other processes

An overview of water treatment is given in Chapter 3 of the Set Book, covering traditional, modern and future technologies. Catchment control as a means of ensuring good quality water for abstraction was the norm in the past but stringent drinking water quality requirements mean that it is no longer viable as a single line of defence.

The 1980 EU Directive on Drinking Water led to the introduction of modern treatment techniques (such as dissolved air flotation and granular activated carbon adsorption) in order to meet new water quality standards, principally related to organic compounds.

Table 3.1 in the Set Book offers a means of selecting possible treatment processes for particular pollutants, but the point is made that other considerations, apart from technique, are important. These include the financial resources obtainable, the amount of space available, the availability of skilled labour, etc. It may be that modification at the intake point will render the raw water less difficult to treat (for example, if the water was pumped from a well by the river, after being filtered through the sand, instead of being drawn directly from the river).

Typical process trains for lowland waters are given. Note the fine minute-aperture filters for removal of any *Cryptosporidium*.

Groundwater is usually simpler to treat than surface water, with iron and manganese often being the only major pollutants present. A flow diagram to treat such waters is given. More recently, groundwaters associated with high-rate surface water recharge mechanisms, such as fissured limestone aquifers, have been recognised as being more prone to pollution than previously thought, depending on local land use patterns. *Cryptosporidium* is regarded as a possible high risk at such sites, hence the need for more stringent treatment facilities to be included in the process trains for such vulnerable sources.

The prime concern as regards water supply in developing countries is the production of water that is microbiologically safe. Some 25 000 people, mostly infants, are said to die worldwide each day from consuming unsafe water. It is best to use (usually) safe sources, such as groundwater. Failing this, a relatively safe surface water source should be sought. In all cases, disinfection is a must, to ensure water-borne pathogens are eliminated. The water should also be aesthetically acceptable.

The World Health Organization encourages the concept of multiple barriers with regard to pathogen control. This is achieved by having several stages in the treatment process, such that failure in one doesn't render the water immediately unsafe.

Looking to the future, lead, *Cryptosporidium* and disinfection by-products are identified as major concerns. The means used to reduce lead in water supplies are discussed. In particular, the problem of pH is highlighted. The lowering of the level for lead in water is of particular concern to water suppliers. Treatment is being modified for this purpose but it must be noted that the lead is often leached from the pipework that connects consumers' properties with the water mains. In many instances it is the customer's pipework that may be the major contributing factor to higher-than-desirable lead levels in supplies to individual properties. *Cryptosporidium* removal by membranes is mentioned. Membranes are also ideal for removing organics from peaty waters. These would otherwise form trihalomethanes upon disinfection with chlorine. Other disinfection by-products are discussed, such as bromates formed as a result of ozonation of bromide-containing waters. There is a suggestion that the use of chlorine itself could be restricted. Indeed this is the case already in the Netherlands.

The author concludes Chapter 3 by postulating that the greatest changes are likely to be in monitoring and control systems for water intake points and water distribution networks. These will allow water treatment to be optimised with regard to changing raw water quality, and ensure that good water quality is maintained all the way from treatment plants to homes.

1.10.1   General principles of water treatment

As stated earlier, the prime objective of water treatment for potable supply is to produce a wholesome water for human consumption. The treatment operations involve the removal of gross solids, turbidity and colour, iron, manganese, aluminium, and various trace compounds such as residual pesticides, if these are present. Most of these parameters were considered in T210/T237 *Environmental Control and Public Health*. Figure 2 shows a possible sequence of operations for the treatment of surface (river or lake) water for potable supply.

Solids removal

Gross solids such as floating vegetation (e.g. branches and twigs) are removed by coarse and fine screens at the intakes to water treatment works. The next parameter that requires consideration is turbidity, which is undesirable both for aesthetic reasons and because it impairs the efficiency of disinfection further on in the treatment process. Turbidity is caused by the presence of finely divided, sometimes colloidal, suspended matter such as silt, decaying vegetation and algae. It can be reduced by simple sedimentation in reservoirs or settlement tanks. The remainder may be removed by filtration through sand filters, or it may be necessary to use coagulation and flocculation, followed by sedimentation or flotation, and then filtration.

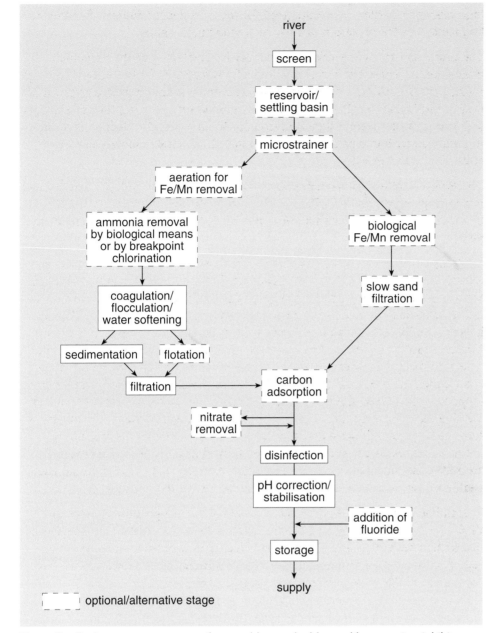

Figure 2   Surface water treatment for potable supply. Note: pH correction inhibits corrosion in the mains distribution system

Microstrainers may be employed prior to chemical coagulation or sand filtration. They are especially useful for removing algae.

SAQ 8

Differentiate between coagulation and flocculation. (Refer back to T210/T237 *Environmental Control and Public Health* if you are unsure.)

Removal of colour

Colour in natural waters is due to the presence of organic matter derived from decaying plant material. Chemical coagulation is used to remove colour, where the colour compounds are precipitated and/or adsorbed onto surface flocs. The Sirofloc^{TM} process (see Section 4.2.6) is also effective for removal of colour. Ozonation is another option – the ozone breaks down the colour-conferring compounds (e.g. the humic and fulvic acids from decaying plant material) into substances that are easily biodegraded.

Removal of iron, manganese and aluminium

Iron, manganese and aluminium can be leached from soils by acidic or low pH waters and can be present in solution, or as complexes with organic compounds in water. This is especially so with underground waters. When the dissolved oxygen content of an underground water is negligible (i.e. it is an anoxic water) iron is often present in the soluble ferrous form.

Iron and manganese contribute taste to water, and can stain laundry and cutlery. The staining with manganese is much more severe. The usual procedure for removal of these contaminants is to oxidise the iron and manganese, found in soluble form as ferrous (Fe^{2+}) and manganous (Mn^{2+}), to their insoluble forms, these being the ferric (Fe^{3+}) and manganic (Mn^{4+}) forms. The most common oxidising agents are oxygen and chlorine. Other oxidants, such as ozone, chlorine dioxide, and potassium permanganate, have also been used. The oxidation of iron is dependent on pH and temperature.

The oxidation of manganese is slower than that of iron, and is again dependent on pH. The oxidation is catalysed by manganese dioxide.

In instances where there are concentrations of iron and manganese, two-stage treatment – first removing iron and then manganese – is common.

Aluminium and iron can be precipitated and removed under the same conditions required for colour removal by chemical coagulation. Manganese remains soluble under these conditions. Effective precipitation of manganese can be achieved by a combination of increased pH ($\geqslant 9$) and the addition of an oxidising agent such as chlorine.

Simple aeration can also be used to precipitate iron as its hydroxide. Aeration may be by various means (see the Set Book, Chapter 4: Preliminary treatment, section 'Types of aerator'). Manganese usually needs an oxidising agent, such as chlorine, to form an insoluble hydroxide to aid its removal to comply with the drinking water quality regulations. Ion exchange can also be used to remove iron and manganese.

Biological methods for the removal of iron and manganese have been developed. In the Vyredox^{TM} process, iron and manganese are removed *in situ* within an aquifer by a mechanism thought to be partially microbiological, initiated by the injection of aerated water to establish a zone of oxidation and precipitation around the abstraction point.

Other processes also using micro-organisms have been developed. One, by Degrémont, uses bacteria in sand filters (Figure 3) to remove iron and manganese. The system utilises mainly *Gallionella ferruginea* and species from the *Leptothrix* and *Sphaerotilus* families (which are present in raw waters and can multiply on sand filters under appropriate conditions) to remove iron by oxidation to the Fe^{3+} state which precipitates out of solution. *Pseudomonas manganoxidans* and species from the *Leptothrix* and other genera are used to remove manganese.

The biological removal of manganese is more difficult to control than the biological removal of iron as the activity of the manganese bacteria is strongly dependent on correct pH and dissolved oxygen levels. The efficiency of the biomass is lower at acidic pH values and the time required to establish the biomass is much greater.

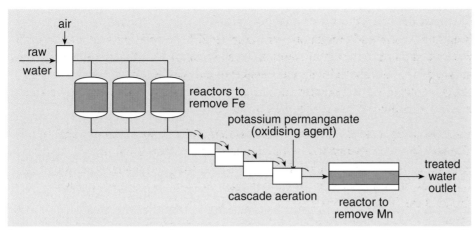

Figure 3 Biological removal of iron and manganese [Courtesy of Degrémont, UK, Ltd]

The mechanisms of bacterial oxidation of the iron and manganese are mainly by:

(a) primary extracellular oxidation by enzymatic action, and

(b) secondary extracellular oxidation caused by the catalytic action of polymers excreted by the bacteria.

The advantages claimed for the biological process (over the chemical option) are:

(a) the greater removal of iron and manganese;

(b) the higher rate of removal of the ions, resulting in a more compact plant being feasible;

(c) the ease of dewatering of the sludge produced.

However, it is stressed that not all raw waters can be economically treated to produce an environment which is appropriate for bacterial activity, so pilot studies are essential before a final choice is made.

SAQ 9

In the treatment of water for potable supply, list the possible origins and means of removal of the following pollutants:

gross solids, turbidity, colour, iron, manganese and aluminium.

Removal of trace organics

There has been increased interest in the removal of trace organics from drinking water. Trace organic compounds, such as pesticides, are of concern due to the health risks they pose. Other compounds which are biodegradable (referred to as assimilable organic carbon, or AOC) are of interest because of their ability to promote bacterial growth in distribution systems, leading to a deterioration in potable water quality from the treatment works to the point where the water is delivered to the consumer. The main trace organics of concern are:

■ pesticides and herbicides

■ volatile organic compounds (VOCs)

■ trihalomethanes (THMs)

Pesticides   Traces of pesticide can be present in surface and groundwater due to widespread and prolonged agricultural use.

The EU has set a limit of 0.1 μg l^{-1} for individual pesticides (except for aldrin, dieldrin, heptachlor and heptachlor epoxide whose limits are 0.030 μg l^{-1}).

The most common means of reducing the trace organic content in water is to use granular activated carbon (GAC). The GAC is used in a structure similar to that of a rapid gravity filter, but the sand is replaced by activated carbon.

Adsorption onto activated carbon is a physical process whereby molecules become attached to the activated carbon principally through van der Waal's forces (weak attractive forces between molecules due to fluctuations in the electronic configuration of the molecules).

A wide variety of carbonaceous materials has been used to produce activated carbon. These include bituminous coal, peat, wood and coconut shells. Specific surface areas of $600–1500$ m^2 g^{-1} have been achieved. Large molecules are adsorbed faster than small ones, and the rate of adsorption increases with temperature. The rate of adsorption is proportional to the square root of the time of contact of the adsorbate (the molecules to be adsorbed) with the activated carbon in the process train.

Adsorption can be modelled using isotherms (such as the Langmuir or Freundlich isotherms, described in Chapter 11 of the Set Book), and these can then be used to predict the mass of activated carbon needed to adsorb a given mass of a specific pollutant. The Freundlich isotherm is the more widely used of the two. As pollutants are rarely present alone in water, the actual quantity of activated carbon required in a particular situation is likely to be greater than that estimated using isotherms alone. Minimum contact times for GAC beds should be 5–20 minutes when the medium is clean. For maximum run-times, the water to be treated by activated carbon should have a very low turbidity (nominally less than 0.2 NTU).

Eventually the carbon becomes exhausted and has to be replaced or regenerated in a high temperature furnace. Note that it is adsorption, rather than absorption, with activated carbon.

Thames Water has devised a system whereby activated carbon sandwiched between conventional slow sand filters is used to adsorb pesticides. Micro-organisms present within the pores of the activated carbon consume the molecules of pesticide trapped within the pores. The system is referred to as biological activated carbon (BAC). Biological activity is encouraged by the use of ozone to break the complex pesticide molecules into simpler, biodegradable components.

Some installations use hydrogen peroxide together with ozone (termed 'advanced oxidation') as a method of increasing the efficiency of the oxidation process.

The treated water has an extra benefit in that it has a lower total organic carbon (TOC) content after passage through the BAC, and this results in less chlorine being needed for disinfection. The lower TOC level also results in a reduction in trihalomethane formation.

Volatile organic compounds   These are typically industrial, chlorinated solvents which find their way into groundwater. Occasionally surface waters are also affected. They can be removed by air stripping or by adsorption onto activated carbon. Air stripping involves passing air through a packed column countercurrent to the flow of contaminated water. The packing offers a high surface area for the water to establish contact, thus facilitating the transfer of the contaminant from the water to the air.

Trihalomethanes (THMs)   Trichloromethanes are a group of compounds with a general formula CHX_3 where X may be any halogen (Cl, Br, F, I, or a combination of these). Long-term, low-level exposure to trichloromethanes can lead to rectal, intestinal or bladder cancer. They occur in water supplies as a by-product of chlorination when chlorine reacts with organic matter present in the water. The amount of trihalomethanes (THMs) produced is proportional to the concentration of organic matter present. The THMs of most concern are:

- trichloromethane (chloroform), $CHCl_3$
- tribromomethane (bromoform), $CHBr_3$
- bromodichloromethane, $CHBrCl_2$
- chlorodibromomethane, $CHBr_2Cl$.

Once formed, they are extremely difficult to remove. Methods of reducing THM formation include:

- removal of the THM precursors by chemical coagulation prior to disinfection with chlorine
- use of activated carbon to remove THM precursors, or the THMs themselves after disinfection
- control of the chlorine dose to inhibit THM formation in the first place
- use of alternative disinfectants to chlorine, such as chlorine dioxide or ozone.

The limit of THMs in drinking water is set at 100 $\mu g\ l^{-1}$. The value is based on the sum of the concentrations of chloroform, bromoform, dibromochloromethane and bromodichloromethane.

SAQ 10

(a) Give two reasons why trace organics are of concern in drinking water.

(b) List the systems usually adopted for the removal of the three main groups of trace organics encountered in water abstracted for potable supply.

(c) What is the limit set by the EU for individual pesticides in drinking water?

Cryptosporidium and Giardia

Most water-borne pathogenic organisms are eliminated in the disinfection stage of potable water treatment. However, some, such as the protozoan parasites *Cryptosporidium* and *Giardia*, are not.

Cryptosporidium can block the body's ability to digest nutrients, cause severe diarrhoea and even kill – more than 100 people died in an outbreak of cryptosporidiosis in the USA in 1993. A major outbreak in the UK (in the Oxford/Swindon area) in 1989 prompted a committee (the Badenoch Committee) to be set up to study the whole subject. In 1995, some 355 people contracted cryptosporidiosis in the Torbay area of Devon. This was suspected to be due to inadequate treatment of the drinking water. The water company responsible (South West Water) was prosecuted but the judge at Bristol Crown Court threw the case out due to a lack of suitable admissible evidence, and adjudged South West Water to be 'not guilty'. This led to the DWI introducing a more stringent monitoring system to check water quality at water treatment plants in high-risk areas, on a continuous 24-hour basis, 365 days a year.

The *Cryptosporidium* oocyst (the stage of the parasite which can survive outside the host) gets into water via sewage from infected people, or from slurries originating from infected animals. The oocyst is round with a diameter of 4–7 μm.

The parasite can occur in low numbers in water but has a very low infective dose – one oocyst is sufficient to cause the disease!

The parasite is very resistant to chemical attack and the most common means of preventing it getting into the treated supply at present is by filtration using membranes. Slow sand filters and rapid gravity filters (with coagulation) are also used.

An expandable bed filter (Fibrotex®) developed by Kalsep is being used in south-west England for the removal of *Cryptosporidium* oocysts from sand filter backwashings.

The filter element comprises thousands of crimped nylon fibres arranged in a bundle around a central core. During filtration, the fibre bundle is twisted and compressed to form a tight helical matrix. The feedwater passes radially from the outside to the inside of the element, with oocysts being captured within the fibrous matrix and held until a preset loading is reached.

Once the filter element is fully loaded, a vacuum steam pasteurisation process inactivates the oocysts. Five minutes at 60 °C is required for this, after which the filter is backwashed with water. The filter element is first untwisted and stretched and is then 'wrung out' by an alternate twisting motion. A small backwash flow of water is introduced and this is drawn into the individual fibres and expelled, along with the oocysts, as the fibres are alternately stretched, squeezed and relaxed. Removal rates of better than 99.9% are claimed. The cleaned backwash water is recycled.

Other systems utilising ultrasound and UV radiation have been tested for the elimination of *Cryptosporidium.*

Wound-fibre filters have been used for *Cryptosporidium* removal, achieving elimination rates of about 99.5%. The filters typically consist of polypropylene fibres wound round a polypropylene core, producing a nominal pore size of 1 μm. The filters can be installed in household plumbing systems for the elimination of *Cryptosporidium.*

The Sirofloc™ process has also been considered for oocyst removal. The magnetite becomes positively charged in acidic waters and abstracts the negatively charged oocysts. Tests using this process have indicated removal efficiencies of 99.9%.

Research has also been focused on traditional disinfectants such as chlorine, ozone, chlorine dioxide (ClO_2) and monochloramine. Ozone was found to be the best, with a 90% kill at high concentration, followed by chlorine dioxide, free chlorine and monochloramine. The studies have suggested that oocysts die more easily if they are exposed to ozone at higher temperatures and lower pH values, but the high dosages of ozone required for an effective kill may make such exposure impractical.

Chlorine dioxide is a strong oxidant produced by mixing a solution of chlorine gas with sodium chlorite solution in a packed reactor:

$$Cl_2 + 2NaClO_2 \rightarrow 2NaCl + 2ClO_2$$

It is a stronger oxidising agent than chlorine and is never stored or used as a gas but is produced on-site as a liquid, on demand.

Giardia is another parasite originating from animals (mainly beavers) that can contaminate watercourses and lakes. It also causes severe diarrhoea. *Giardia* cysts (oval, 8–14 μm long and 7–10 μm wide) are said to be destroyed by a 2 mg l^{-1} residual of chlorine after 10 minutes contact time at pH 6–7, or a 3 mg l^{-1} residual at pH 8.

The cysts can also be removed, to a significant extent, by slow sand filtration, or rapid sand filtration following coagulation treatment.

Ozone and chlorine dioxide have both been found to be very effective against *Giardia* cysts. Boiling the water for 20 minutes is also said to destroy the cysts. In household supplies, wound-fibre filters can be very effective against *Giardia.*

SAQ 11

(a) Which is the most commonly used method of eliminating *Cryptosporidium*?

(b) Give the names and chemical formulae of two oxidising agents which have been found to be very effective against *Cryptosporidium* and *Giardia*.

Algal toxins

Cases have arisen where blue-green algae have flourished in reservoirs, releasing toxins. Animals, especially dogs, have been affected after drinking water containing blue-green scum, or by swimming in algal-laden shallow waters, thus exposing their bodies to the blue-green scum. In terms of treatment for potable supply, algae can be removed by sand filtration. Any residual toxin in the water can be eliminated by adsorption onto activated carbon. Dosing filtered water with oxidising agents such as potassium manganese (VII) (potassium permanganate) or ozone is also effective in removing or degrading the toxins. Chlorine, however, has been found to be ineffective except at low pH levels.

Miscellaneous components

Other components of treated water that may require attention are ammonia, lead, nitrates, hardness, taste and odour, and arsenic.

Ammonia removal   Ammonia removal for wastewaters was covered in T210/T237 *Environmental Control and Public Health*. The methods commonly used in water treatment are:

- air stripping
- breakpoint chlorination.

Lead control   Lead levels in drinking water have to be reduced to 10 μg l^{-1} by December 2013. The most economical way of achieving this in systems where there is lead piping is to dose the water with orthophosphate. This results in insoluble lead phosphate being formed. The lead concentration in the water can be reduced to about 5 μg l^{-1} by this means. The pH has to be controlled to achieve this. For low-alkalinity waters, the appropriate pH range is 7.5–8.3.

Nitrate removal   Nitrate removal was also covered in T210/T237 *Environmental Control and Public Health*.

Formula feed prepared with drinking water containing high levels of nitrate can lead to methaemoglobinaemia in babies. As a result there is a limit on nitrates in drinking water supplies.

The easiest option for dealing with a source which has high nitrates is to blend it with one which is low in nitrate. This, however, may not always be practicable, and one of the following process options (see also Section 1.11) may have to be employed:

- ion exchange
- biological treatment
- reverse osmosis
- electrodialysis.

Ion exchange is the simplest of the above processes. A special resin attracts the nitrate ion, and replaces it with a chloride ion. This makes the product water more corrosive than the incoming feed. The regeneration of the resin produces a highly saline waste product that has to be disposed of.

While most ion exchange systems use synthetic resins, many are naturally occurring, such as the zeolites. The biggest use for ion exchange is in the softening of water, but it has been used for the removal of chromium, barium, strontium and radium.

Hardness   Hardness in water is related to the tendency to form scale, and results almost entirely from calcium and magnesium compounds. Removal of hardness (or water softening) can be undertaken by ion exchange or chemical precipitation (see Set Book, Chapter 11: Other processes, section on 'Water softening').

Taste and odour   Taste and odour are often due to synthetic organic compounds or naturally occurring metabolites present in the water source. They can be removed using granular activated carbon or ozone.

Arsenic   Arsenic is widely distributed in the Earth's crust (it makes up some 0.00005%). It is used in the production of alloying agents, insecticides and herbicides, wood preservatives, pigments and lead shot. It is introduced into water by the dissolution of ores, from the discharge of industrial effluents, and from atmospheric deposition. Inorganic arsenic is a human carcinogen, and a relatively high incidence of skin cancer has been observed in populations (e.g. in Bangladesh) ingesting water containing high concentrations of arsenic (e.g. 300 μg l^{-1}). The limit in drinking water under the 1998 EU Drinking Water Directive is 10 μg l^{-1}. The common forms of arsenic found in groundwater are arsenate (e.g. H_3AsO_4, H_2AsO_4, AsO_4^{3-}) and arsenite (e.g. H_3AsO_3, H_2AsO_3, AsO_3^{3-}). Of these two forms, arsenite is the more toxic, and is associated with anaerobic groundwaters, whilst arsenate is found in aerobic waters. Arsenate is more easily removed than arsenite, and hence it is beneficial to oxidise any arsenite before undertaking removal.

One of the methods employed is coagulation and flocculation using aluminium and/or ferric salts. Arsenic is incorporated into the hydroxide sludge.

Another method is direct filtration using aluminium or ferric salts. This involves small quantities of the coagulant being added to the contaminated water, and the water then being filtered using rapid gravity filters.

Arsenic can also be adsorbed onto specialist media such as activated alumina, bone char, manganese oxide, and granular ferric hydroxide. Other processes such as ion exchange, electrodialysis, co-precipitation onto microfiltration membranes, nanofiltration and reverse osmosis, have also been applied.

Addition of fluoride

In certain areas of the country, fluoride is added to the water supply in order to help prevent tooth decay (see T210/T237 *Environmental Control and Public Health*).

Disinfection

Finally, the water is disinfected (usually by chlorine) to eliminate all pathogens before being sent into the water supply system.

SAQ 12

Write short notes on the following:

(a) Blue-green algae.

(b) Breakpoint chlorination.

(c) Biological fluidised bed.

1.11   Membrane processes

> **READ**
>
> Set book Chapter 10: Membrane processes

You were introduced to membranes in T210/T237 *Environmental Control and Public Health*. They are an effective way of eliminating tiny particles (even ions) from water. The pore sizes of the different types of membranes are not defined strictly, so different pore sizes may be quoted for similar membranes. The different membrane systems can be listed (in increasing pore size):

- reverse osmosis
- nanofiltration
- ultrafiltration
- microfiltration.

The two most common configurations of membranes, spiral wound and hollow fibre, are described on p.173 of the Set Book.

In order to prevent premature blockage, the water to be passed through the membrane has to be pretreated. Suspended solids are removed by filtration or coagulation/flocculation. Ultrafiltration membranes may be utilised prior to reverse osmosis. Scaling is prevented by pH adjustment, or by dosing with chemicals.

Drinking water produced by reverse osmosis (say, from seawater) is aggressive, due to a lack of alkalinity. A common treatment to remedy this is to pass it through a column of limestone. The major disadvantage associated with reverse osmosis is the high cost due to its power requirement.

Nanofiltration membranes have found application in nitrate removal, removal of organic matter which would serve as precursors for trihalomethanes, and in water softening. Because their pore size is larger than that of reverse osmosis membranes, they are not used for desalination.

Ultrafiltration membranes have larger pore sizes than nanofiltration systems. They have been used for protection against *Cryptosporidium* in water treatment plants. Their energy requirement is less than reverse osmosis membranes, because they operate at lower pressures.

Ultrafiltration is a cross-flow process. Here, the water requiring treatment flows parallel to the membrane. An advantage of this mode of operation is that the fouling of the membrane is less, since the flow of water inhibits the deposition of solids. Ultrafiltration is another process that is used for *Cryptosporidium* removal in water treatment.

Microfiltration has a larger pore size than ultrafiltration.

Electrodialysis is another membrane process, which uses an electric field to separate ions of different charges. Ion-exchange membranes with fixed charge groups attract mobile ions with the opposite charge from the water. The membranes are arranged as cation and anion exchange membranes in cells between the anode and cathode (see T210/T237 *Environmental Control and Public Health*). Anionic membranes are permeable to anions, and impermeable to cations, while the opposite is the case with cationic membranes. When an electric current is applied, cations attempt to move towards the cathode, while the anions attempt to reach the anode. The ion-selective membranes result in the cations and anions being trapped in altering cells of ion-enriched solution and ion-depleted solution.

1.12   Treatment of water from underground sources

In general, groundwaters are less polluted than surface waters and very often they are almost clean. Table 12 compares the two. Many of the mineral waters available in supermarkets have been bottled without any treatment whatsoever. For general water supply, however, groundwaters, if clean, are disinfected before distribution. This is to safeguard against microbiological contamination while the water is in the distribution network.

Table 12   Characteristics of ground and surface waters

Groundwater	Surface water
Constant composition	Varying composition
High mineralisation	Moderate to low mineralisation
Low turbidity	High turbidity
Little or no colour	Colour
Usually bacteriologically safe	Micro-organisms present
Minimal dissolved oxygen	Contains dissolved oxygen
High hardness	Low hardness
Can contain hydrogen sulphide, iron and manganese	Tastes and odours likely
	Possible chemical toxicity

Certain groundwaters may contain high levels of iron and manganese and these are removed using one of the methods described earlier. Other aspects of groundwater treatment may involve units that remove nitrates or hardness.

Figure 4 shows a schematic of the possible routes for the treatment of groundwater for potable supply.

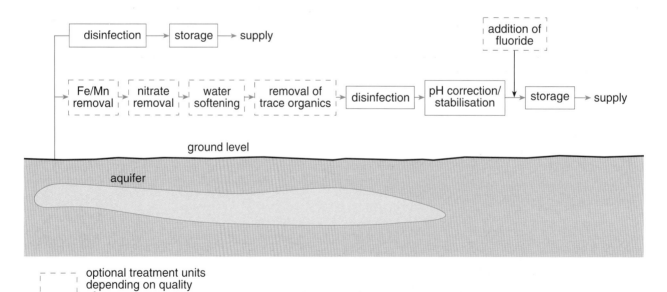

Figure 4   The various treatment options for groundwaters

Instances have occurred where solvents have entered groundwaters, requiring activated carbon to be used in the treatment phase to remove these organic compounds.

1.13 Similarities and differences between water treatment and sewage treatment

In studying the treatment of water for potable use, you will have noticed that many of the unit operations seem to be similar to those encountered in sewage treatment (see T210/T237 *Environmental Control and Public Health*). It is easy to get muddled between the two purification processes. Potable water treatment is mainly chemical based, for the processes of coagulation, flocculation, filtration and disinfection, whilst sewage (or effluent) treatment is mainly biologically based, using micro-organisms to degrade pollutants in the wastewater. In potable water treatment all the micro-organisms are destroyed but in sewage treatment some can remain in the final discharge.

Water abstracted for potable water treatment will usually be only mildly polluted. The treatment steps, in the main, will involve the removal of gross solids (branches, twigs, etc.), suspended and colloidal solids, colour, and ions of iron and manganese. The water will be disinfected to eliminate any micro-organisms.

With sewage, the major pollutants will be carbonaceous matter (both suspended and dissolved) and grit (where the sewerage system is a combined system). So oxidation of the carbonaceous matter is a vital part of sewage treatment and this is usually carried out by micro-organisms using systems such as activated sludge or biological filters.

If after secondary treatment of the sewage, the levels of suspended solids, plant nutrients (such as nitrate and phosphate) or ammonia are still unacceptable for discharge, tertiary treatment can be carried out. Table 13 lists the main methods available for elimination of these substances. You should refer to T210/T237 *Environmental Control and Public Health* if you are unsure about any of these processes, some of which (such as the use of grass plots for suspended solids removal) are only suitable for sewage treatment and cannot be used for water treatment. Figure 5 summarises the main stages in the treatment of domestic sewage.

Table 13 Possible methods of removal of various substances in tertiary sewage treatment

Contaminant	Methods of removal
Suspended solids	microstraining
	sand filtration
	further settlement in lagoons, etc.
	grass plots
	reed beds
Nitrate	ion exchange
	fluidised bed reactors
	membranes
	electrodialysis
Phosphate	precipitation using lime, alum, ferric chloride or ferric sulphate
	reed beds
	application of anaerobic and aerobic conditions in an activated sludge unit
Ammonia	biological nitrification/denitrification
	air stripping (or steam stripping where steam can be used in place of air)
	breakpoint chlorination
	ion exchange

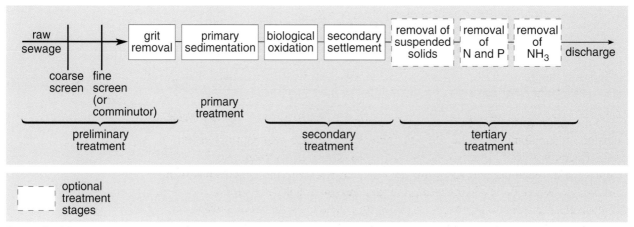

Figure 5   Unit operations and processes commonly used in sewage treatment

SAQ 13

Identify from the following list of treatment options those that are inappropriate for use in potable water treatment.

(a) coarse screens

(b) fine screens

(c) grit removal

(d) microstrainers

(e) coagulation and flocculation

(f) sedimentation

(g) flotation

(h) filtration

(i) reed beds

(j) ion exchange

(k) fluidised beds

(l) membranes

(m) air stripping

(n) breakpoint chlorination

(o) electrodialysis

(p) grass plots.

2 THE WATER FRAMEWORK DIRECTIVE

2.1 Introduction

The publication of the Water Framework Directive (2000/60/EC – see the T308 Legislation Supplement) in the Official Journal of the European Communities on 22 December 2000, marked the end of a prolonged series of ministerial seminars, and technical and scientific reviews that followed the conclusions of a Community Water Policy Ministerial Seminar held in Frankfurt in 1988, emphasising the need for Community legislation to include the ecological quality of water resources. However, it was not until 1996 that the European Commission was formally requested to develop a Council directive that firmly established a framework for a cohesive European water policy amongst Member States, utilising many of the existing water quality directives.

Implementation of the directive poses considerable varied and complex technical challenges to Member States. So at European level, a number of Common Implementation Strategy Working Groups comprising scientific and technical specialists from across the European Union have been busy providing guidance to Member States on the implementation of the general aspects of this directive. This guidance is aimed at introducing best possible practice in complying with the terms of the directive. It will also enable effective comparisons to be made by the European Commission on the progress made by Member States in implementing and complying with the terms of this directive.

The directive has replaced many of the earlier water quality directives introduced by the European Community in the 1970s. These include:

- Directive 75/440/EEC – concerning the quality required of surface water intended for the abstraction of drinking water
- Directive 79/869/EEC – concerning the methods of measurement and frequencies of sampling and analysis of surface water intended for the abstraction of drinking water
- Directive 78/659/EEC – on the quality of freshwaters needing protection or improvement in order to support fish life
- Directive 79/923/EEC – on the quality required of shellfish waters
- Directive 80/68/EEC – on the protection of groundwater against pollution caused by certain dangerous substances.

Article 22 of the Water Framework Directive deals with the timescales for the repeal of earlier water quality directives and in particular indicates transitional arrangements in respect of Directive 76/464/EEC – the Dangerous Substances Directive. The fundamental objective of this directive is to minimise and, where possible, prevent deterioration of the water quality and ecological status of surface waters, including marine waters, by pollution. It also seeks to minimise further deterioration of groundwater quality, with the justifiable aim of achieving a 'good' status for all waters, 'good' being variously described in the wide range of parametric standards and conditions set out in the many Annexes to the directive.

There will be a significant impact on the management of both surface and groundwater resources, and associated water quality; together with fisheries, conservation, environmental monitoring, flood alleviation and planning, to meet future socio-economic requirements and targets. Many Member States will be significantly challenged by the terms of this directive, following its transposition into domestic legislation imposed by the European Union. Even in the United

Kingdom there are significant differences in the way the water environment will be managed in the future. For example, England and Wales have traditionally applied a comprehensive licensed system of control of abstractions, with due regard for aspects of quality and quantity, thus influencing environmental impact of individual abstractions; in Scotland, most major abstractions for public water supplies and hydro-power schemes are authorised by individual historical Parliamentary Acts and Orders. Not surprisingly, conditions set at the time such authorisations were enacted were based on local negotiation rather than an appreciation of environmental need. By contrast, in Northern Ireland, abstractions from surface water and groundwater resources have historically not been subject to licensed conditions. There may be a few isolated abstractions where locally agreed conditions between water assets are respected, but these are informal arrangements only. Thus, there are significant environmental and economic factors to consider for the future.

Obviously, as the skills required to consider risks relating to environmental impact have developed alongside the increasingly wider appreciation of the value of the environment to the quality of life now demanded by society, a review of current abstraction conditions is likely to lead to proposals for change. In Scotland, and probably many other similar regions of the European Union, there will be no easy mechanism available to accomplish this, This is in direct contrast to England and Wales where such conditions have been in place and subject to regular monitoring patterns in the past.

The implications of this directive are wide and far reaching. It establishes a progressive and integrated approach to the protection, improvement and sustainable use of all water resources – rivers, lakes, reservoirs, groundwaters, coastal and transitional waters (estuaries and lagoons), integrating both quality and quantity requirements.

2.2   Key factors

Article 1 of the directive clearly highlights its purpose – to provide a framework of environmental objectives that will:

- prevent further deterioration of water resources
- protect and enhance the status of aquatic ecosystems
- take account of the impact on wetlands of the management of water quality and quantity
- progressively enhance the protection and improvement of the aquatic environment by reducing and where necessary, phasing out discharges and emissions of hazardous and environmentally detrimental substances
- promote sustainable water consumption and general use by communities and businesses
- help to minimise deterioration and further pollution of groundwater resources
- assist in mitigating the impact and effect of droughts and floods.

Article 4 of the directive outlines the overriding environmental objectives. These include consideration of the demands that are likely to be placed on individual aquatic systems, dealing specifically with:

- flow regulation and transfers of flow regimes with particular regard to potential damage to habitats and associated terrestrial ecosystems
- abstraction from and return to an aquatic system (cooling waters)
- pollution – point source and diffuse discharges and drainage
- emergency management following incidents that may be natural disasters or as a consequence of accidents or subversive activities.

Not surprisingly, the emphasis will be on ecological, rather than simply chemical quality. This is a significant departure from most historical approaches to pollution prevention activities and legislative/regulatory requirements. Ecological status is subdivided into five broad categories – high, good, moderate, poor and bad. This status broadly comprises assessments or measurements of biological, hydromorphological and physico-chemical characteristics. Obviously, biological aspects include aquatic flora and fauna, invertebrates and fish population, and is regarded as being of paramount importance. The remaining characteristics are regarded as supportive only in maintaining a satisfactory biological status. Before a water mass or body can be classified as having a 'good' status, both the ecological and chemical status must also be 'good'.

Undoubtedly, the interpretation of 'good' status, largely described in the various Annexes to the directive will provoke considerable debate amongst the Regulators and those stakeholders with vested and specific interests in the environment and the well-being of society. For the practitioners, a consistent working definition or at best a well-honed description, will be a most welcome aspect to aid a more general understanding of what is needed to achieve 'good' status.

A significant aspect of the directive is the requirement of Member States to manage effectively all waters within their geographical boundaries in order to achieve good ecological status. Several Member States will need to work closely together to move towards attaining this objective, especially where some rivers, such as the River Rhine, cross international boundaries, are navigable and consequently are major transportation arteries impacting on the well-being of the economies of these countries. The risk of cargoes directly damaging the ecology of such 'waterways' must constantly be addressed with the relevant stakeholders.

SAQ 14

Explain the prime objective of the Water Framework Directive.

For some Member States, previous approaches to their systems of water management have been much less comprehensive than that now demanded by the directive. This is the case in parts of the United Kingdom, where England and Wales have traditionally been more focused on the catchment area than Scotland, which was more sectorally based. The principles of the directive appear to be quite straightforward, but the real challenges emerge from the implementation of the Technical Annexes that accompany the directive. These are complex, but provide the fundamental principles that establish the scope and intent of the administrative procedures and regulatory provisions that Member States will need to include in their internal legislation. Promoting the sustainability of the entire water environment both above and below ground is a major objective of this directive; hence the need to ensure that the implementation of the Technical Annexes is consistent throughout the European Community. It is reassuring to note that the directive introduces a standard approach to plan ahead to help ensure the protection, improvement and continued beneficial use of the European Community's water environment. Technical Annexes II and V provide the core of this approach.

Previously, relevant European water legislation had imposed specific objectives to protect certain users of the water environment from the effects of pollution, with particular emphasis on restrictions regarding environmentally damaging chemicals. This approach continues and is included in the directive's requirements for specific protected areas and the priority substances (currently 32 in number, see the T308 Legislation Supplement) that will be listed in Annex X to the directive. These priority substances are mostly organic chemicals that are persistent in, and damaging to, the environment. They include pesticides, herbicides, polyaromatic hydrocarbons (PAHs), certain complex phenolic compounds and chloro-benzenes, tributyl tin compounds (TBTs) and compounds of lead, mercury and nickel.

Protected areas are those water bodies and associated terrestrial areas that have been designated as requiring special protection under European Community legislation, to protect their water quality or the related habitats and species that depend on such water bodies. Article 6 of the directive requires Member States to ensure the establishment of a register of all such areas within four years of the directive coming into force. Types of water bodies to be included in this important register are listed in Technical Annex IV to the directive. These are:

1   Water bodies used for the abstraction of drinking water.

This is a newly designated protected area with the provisions for protection replacing the requirements previously included in the Surface Water Abstraction Directive (75/440/EEC). Article 7 of the Water Framework Directive requires all such areas to be identified and specified as a surface water, groundwater, or, as appropriate, a grouping of both types.

2   Areas designated to protect economically significant aquatic species.

These are mainly protected areas that were established following publication of directive 79/923/EEC to protect shellfish waters.

3   Recreational waters.

These include freshwater and coastal bathing waters as designated by the conditions of the Bathing Water Directive (76/160/EEC), where the standards specified are designed to protect the health of the bathers and other users of these waters. In the United Kingdom, these are mainly coastal bathing waters (beaches), in contrast to other Member States that are wholly or partially land locked.

4   Nutrient-sensitive areas.

These mainly consist of agricultural-related zones that are vulnerable to high nitrate input and will have been designated to meet the terms of the Nitrates Directive (91/676/EEC). The Annex includes those areas regarded as nutrient-sensitive areas when applying the requirements of the Urban Wastewater Treatment Directive (91/271/EEC).

5   Areas designated for the protection of habitats or species where maintenance or improvement of the status of water is important to their protection.

Such areas include all Natura 2000 sites (sites that have been designated under the Habitats and Birds Directives) that have been proposed for inland lakes, river systems, brackish lagoons, estuaries and coastal waters in England and Wales. The Annex encompasses other nominated Natura 2000 habitats and species that are wholly or partly dependent on the quality and sufficiency of the water environment. Examples of this dependence include:

- alluvial alderwoods that rely on sediment transportation and flows in riverine systems
- sand dunes affected by wave action
- marine sediment transportation,
- wetlands – blanket bog, alkaline fens, quaking bog
- groundwater levels
- seabird colonies.

This directive has also introduced quite far-reaching ecological objectives, designed to at least maintain and, where necessary, improve or restore the environmentally beneficial function and social appeal of aquatic ecosystems. The detail of how this might be achieved is contained in the Technical Annexes, where the classification schemes relating to water and ecological quality objectives are specified. Equally importantly, these Annexes ensure that shortfalls in attaining these objectives will also become apparent. However, the economic instruments needed to overcome shortfalls may be extremely onerous and difficult to resolve.

Another major factor in the manner by which the water environment is to be managed in the future by Member States is the introduction of an integrated River Basin Management Planning System. This is fundamental to ensure effective balance of all stakeholders' interests in the controlled use of water resources while maximising environmental benefit. Whereas this will undoubtedly be a novel approach in some Member States, England and Wales have established the catchment management planning approach for well over a decade. By contrast, the history of river basin planning in Scotland is minimal. Where such examples exist in Scotland, as for example in the River Spey, these are of recent origin, with little evidence, as yet, of widespread use. At present in the United Kingdom generally, this approach involves largely voluntary partnerships having been organised to consider local environmentally sensitive issues, so it is unlikely to conform to any national guidance. Nonetheless it is worthwhile recalling that historically in the United Kingdom, national industries that were geographically widespread and under common ownership, such as the coal, steel and china clay industries, sought a common approach to pollution prevention and environmental improvements from a range of regulators applying diverse environmental standards. Some success was achieved with benefit to all concerned.

In England and Wales, the Environment Agency and its predecessor regulatory organisations have already divided rivers and estuarine waters into zones or stretches to assist in the management of point source discharges that may include pollutants. The requirement to establish river basin management planning will also assist in managing other relevant and important contributory aspects such as diffuse pollution and the control of groundwater quality.

This directive will require:

■ all riverine systems and estuaries exceeding 10 square kilometres, and

■ freshwater lakes with surface areas of more than 0.5 square kilometre

to be automatically identified as a water body on its size only. However, there are very many smaller lakes, lagoons and streams which resemble drainage ditches that will also exert adverse impacts on other water masses, and so will require some form of control.

The timescale for providing River Basin Management Plans is based broadly on a six-year planning cycle encompassing four main components:

■ assuring and characterising impacts on river basin districts

■ environmental monitoring

■ setting environmental objectives

■ designing and implementing improvement measures to achieve environmental objectives.

All plans to be published at the start of each six-year cycle are to be preceded by an appropriate period of public consultation. Article 13 of the directive deals with the requirements for River Basin Management Plans, with separate clauses addressing those instances where (a) an international river basin district falls entirely within neighbouring Member States, and (b) where such district boundaries fall beyond the jurisdiction of Member States.

Copious detailed information is required to be included on each individual plan. This is specified in Technical Annex VII and extends to some 35 discrete requirements that must be reported on.

Significantly, subsequent plans in the six-year cycle must also include:

■ summaries of changes that have occurred since the publication of the initial or previous River Basin Management Plan; and also include a summary of the reviews of environmental objectives as required under Article 4 of the directive;

■ an explanation of the progress being made towards achieving the previously agreed environmental objectives, giving reasons for shortfalls. Also, the results of the monitoring programmes should be presented in map form;

■ references to any previous anticipatory measures included in earlier versions of the River Basin Management Plan which were not addressed;

■ a summary of any interim additional measures regarded as being necessary since the publication of the previous version of the River Basin Management Plan. Such a summary is a requirement of Article II (Clause 5) which stipulates that Member States must thoroughly review all possible factors that may contribute to the failure to achieve the specified environmental objectives. Such measures must include evidence of a review of relevant permits and authorisations, and monitoring programmes to reaffirm or amend plans as appropriate.

Article 5 of the directive strongly implies that the approach to river basin planning uses as its core a comprehensive consideration of the impact of human activity – social, recreational, commercial and industrial – on the existing and proposed status of surface and groundwaters. This will, of course, include an economic appraisal of water use.

This assessment must be carried out in accordance with the technical requirements of Technical Annexes II and III to the directive within its initial four-year span, being updated at specified intervals until this directive is amended or repealed. Such a review of the impact of human activity requires a risk assessment of the manner in which environmental objectives and ecological status may be prejudiced by such pressures on water bodies. This will require a sequential process, with inputs from all stakeholders, to the regulatory organisations responsible for formulating River Basin Management Plans on behalf of each Member State. In the United Kingdom, the Environment Agency will carry out this role for England and Wales; and in Scotland the Scottish Environment Protection Agency (SEPA) will be charged with this task.

In Northern Ireland, this task rests with the Northern Ireland Environment and Heritage Service, which, as an Executive Agency of Government, takes the lead in implementing environmental policy in Northern Ireland.

Significantly, in England and Wales, the Environment Agency already has in place a risk assessment outline methodology as part of its Catchment Abstraction Management Strategy for water resources.

Likewise, methodologies have been developed jointly between the Environment Agency, English Nature and the Countryside Council for Wales to assess risks that may not allow the objectives set for Special Areas of Conservation under the terms of the Habitats Directive to be fulfilled. Similar initiatives at this stage are under way in Scotland, with Scottish National Heritage and Scottish Fisheries Research Service and the Scottish Agriculture and Science Agency consulting and supporting the Scottish Environment Protection Agency, as appropriate. In Northern Ireland the Environment and Heritage Service undertakes this task.

It will be important to establish levels of confidence in the accuracy and relevance of the data and information needed from all stakeholders to classify properly the significance of risks to the overall characterisation of water bodies. Obviously, high levels of confidence can be easily achieved without the need for comprehensive risk assessments if a water body is either clearly not at risk, or is so heavily contaminated that it will fail to satisfy the objectives of this directive. This will enable attention to be focused on the more 'doubtful' situations. Historical data will be invaluable in completing this requirement. More definitive assessments will be needed and will emerge by the next due date of 2013, aided by the application of more sophisticated methodologies and data capture systems likely to be developed.

Identifying potential risk factors is a major task and will require dedicated and coordinated approaches. For example, the United Kingdom's Marine Pollution Monitoring Management Group is well advanced in developing a coherent approach to describing and identifying significant risks to coastal and marine waters. The following factors have been specifically identified by the directive:

- water abstraction for water supply, agricultural, industrial and other uses
- point source pollution from all sources and activities
- diffuse pollution from all activities and installations
- flow regulation of surface waters, including water transfers
- land use patterns with particular reference to urban, industrial and agricultural areas, afforestation and transportation
- changes to the morphology of surface water bodies
- artificial recharging of groundwaters for specific purposes such as safeguarding the sufficiency of water supplies
- impact of boat traffic
- flood alleviation and management.

Existing sources of information to aid the risk assessment process include:

- water quality monitoring data – historical and current
- national inventories of abstraction licences, consents to discharge to surface water systems, and river habitat surveys
- flood defence data
- large impounding reservoirs (Reservoirs Act)
- engineering works in transitional (estuarine) and coastal waters (Food and Environment Protection Acts)
- local Authority planning records for land use
- local river catchment and coastal zone management plans
- recently introduced Local Environment Action Plans (LEAPs)
- Catchment Abstraction Management Strategies (CAMS)
- Environmental Statements collated to meet the terms of the Environmental Impact Assessment Directive
- remote sensing studies and surveys such as aerial photography and underwater sonar surveys
- special investigative studies commissioned by public organisations, special committees and research interests
- public Registers of Consents to Discharge, groundwater authorisations and abstraction licences.

To summarise the outline requirements for risk assessments, the continuing programme will essentially be cyclical and interrelated. Requirements will include:

- identifying and reviewing the characteristics of river basins
- assessing impacts from human activities
- implementing action plans to achieve environmental objectives
- designing and prioritising improvement measures and changes to existing practices
- implementing monitoring programmes
- reviewing the results of monitoring programmes
- appraising the impact of improvement measures
- realigning action plans to maintain or accentuate progress towards attaining environmental objectives.

Article 14 of the directive addresses public information and consultation, requiring Member States to be active in involving all organisations, users, industry pressure groups and environmental ecologists in the formulation of River Basin Management Plans. This is essential to demonstrate greater transparency and social involvement in the decision-making process that is designed to improve and sustainably develop the environment for the benefit of all stakeholders. To achieve this, each Member State must publish the schedule and timetable for each River Basin Management Plan at least three years prior to its inception. All relevant issues for each discrete catchment area are to be communicated to relevant stakeholders at least two years in advance, with the initial draft of the 'Plan' being issued one year before its likely enforcement.

Member States will allow at least six months for consultees and the public at large to comment in writing on the 'Plan' and any changes. Thus the duty to inform the public is imperative. Each River Basin Management Plan will highlight the characteristic features of the catchment area, as specified in Technical Annex II to the directive. The influences of human activity on the water concerned and a cogent economic analysis of water usage in the affected area must be clearly focused in the text of the 'Plan'.

SAQ 15

The Water Framework Directive has introduced three significant changes to the current approach to managing the water environment in Member States of the European Community. What are these changes?

2.3   Environmental objectives

The directive subdivides the requirements into three distinct categories in Article 4, namely surface waters, groundwaters and protected areas. Despite this categorisation, the directive specifies three principal requirements which are common to each of these.

- Deterioration of an existing ecological status to be prevented, or at worst, minimised.
- Where necessary, such bodies of water to be protected and restored to enable criteria to be satisfied that will enable protected status to be attained as shown in European Community legislation.
- Water bodies to be restored to good status – or to achieve good potential in the case of heavily modified and artificial bodies, by 2015.

'Heavily modified water bodies' is a term used to describe those parts of a water system that have been physically modified to enable specific uses to be accommodated. Typical examples are flood defence engineering works to facilitate urban regeneration, land drainage, transportation and industrial/hydropower supplies; improving harbour and port facilities; and to make space available for extended residential developments.

For example, in England and Wales over 30 000 km of rivers have flood alleviation facilities and arrangements for their maintenance, with almost 1500 km being used for navigation. Welsh rivers (excluding the upper reaches of the River Severn) support some 50 hydroelectric schemes.

Some physical alterations have tended to create a single water body, whereas others have created more than a single heavily modified water body. Consider the situation where a river flow is interrupted by an impounding reservoir. This will create at least two heavily modified water bodies with differing hydromorphological characteristics. These are, of course, the reservoir and the downstream river (which is subject to modified flow regimes). Not surprisingly, clear criteria need to be developed before meaningful and equitable decisions can be made regarding these types of water bodies. Case studies are being carried out across Europe. Their findings will contribute to the development of consistent guidance for categorisation across Member States.

Artificial water bodies have been created to serve specific purposes and have probably developed definitive aquatic ecosystems. These are mainly canals, some dock areas and impounding reservoirs.

Currently, there is a significant canal restoration programme in progress, with many projects being at the appraisal, planning or active reconstruction stages. Obviously this restoration programme will now have to take account of the need to readjust ecological standards to meet the requirements of this directive.

2.4   Surface water classification

Since the directive has been implemented to improve the overall quality of aquatic ecosystems throughout the European Community, its success will be assessed on the improvements achieved in the aquatic environment and the allied attractiveness of water resources and increased popularity of leisure activities to the general public and water users. Technical Annexes II and V support the provisions of Article 4. For example in England and Wales, and to a lesser extent in Scotland, improvements have been measured historically in terms of the chemical quality of water and the impact of that quality on the aquatic flora and fauna of receiving waters. Nevertheless, it has always been accepted that simply meeting numerical water quality standards is not the finite solution.

For the future, the directive requires a new ecological status classification scheme to be developed for all marine systems, lakes, estuaries and coastal waters. These classification schemes will describe both chemical quality and ecological status, so it is inevitable that the overall status of a surface water body will be determined by the poorest achiever of these factors. At times it may not be feasible technically to achieve 'good' ecological status. Likewise, there may be situations where marginal or major improvements may prove to be disproportionately expensive. In such cases, the directive allows for less stringent measures to be applied.

A vital initial requirement is the identification of criteria to establish the relevant aspects of the chemistry, hydro-morphology, biology and physico-chemical characteristics that would indicate either no changes or inconsequential changes to bodies of water arising from human activities or interactions. These will create 'reference conditions' against which the entire classification scheme can be constructed. These reference conditions will be different for riverine systems, lakes, estuaries and coastal waters. They will also be different for upland and lowland waters such as the relatively oligotrophic Scottish lochs, Welsh uplands, the Lake District in Cumbria and the more eutrophic Norfolk Broads in eastern England.

Similarly for estuarine and marine waters, Member States bordering the Atlantic Ocean and the North Sea will need to use a shared typology for the sake of consistency.

2.5   Groundwater classification

The status of groundwater depends primarily on the impact it exerts on dependent aquatic and terrestrial ecosystems. It is also affected by inputs from surface waters and the level of abstraction from inclusive aquifers. The directive stipulates that Member States introduce a classification scheme to categorise groundwater as being of 'good' or 'poor' status. Those initially regarded as being of 'good' status must be protected from deteriorating to 'poor' status, whereas the latter water bodies must be improved to attain 'good' status by 2015. Given the size and history of certain aquifer systems, this timescale might not be sufficient to reach this goal.

The directive also includes general provisions to protect or limit the ingress of pollutants into groundwater and to take action to reverse known upward trends in the levels of pollutants detected in groundwaters. This effectively requires groundwaters to be protected, especially those used or likely to be considered for use as resources for the abstraction of drinking water supplies. The European Commission has been working to develop proposals for inclusion in a daughter directive on groundwater that will be specific to the prevention and control of groundwater pollution. Of critical importance are the criteria for:

- assessing good chemical status
- identifying significant deteriorating trends in chemical quality
- defining starting points to reverse deterioration in chemical quality.

The directive also includes criteria relating to qualitative matters in its requirement to classify status. This aspect refers to the effects of changes in the level of a groundwater body resulting from abstraction or intrusion (this may also be detected by electrical conductivity or salinity changes). For example, to achieve a good quantitative status, groundwater must not be abstracted at a rate that is greater than the replenishment rate. Such aspects may also modify the chemical composition of groundwaters leading to deleterious impacts on subsequent usage and dependent terrestrial ecosystems. Annex V to the directive requires River Basin Management Plans to display colour-coded groundwater bodies: Green for good status and Red for poor status, the colour coding reflecting both chemical and quantitative status aspects.

SAQ 16

Highlight the major changes introduced in the Water Framework Directive to classify surface and groundwaters.

2.6   Monitoring programmes

Articles 7 and 8 of the directive supported by Technical Annex V relate to the need for Member States to establish monitoring programmes for both surface and groundwater. The data gathered by these programmes will support the formulation of River Basin Management Plans. It will also:

■ contribute to environmental risk assessments;

■ assist in establishing the status of nominated water bodies;

■ help to assess progress in moving towards attaining nominated environmental objectives.

The development of the directive's monitoring programmes will provide an opportunity to review and readjust historical and existing monitoring programmes that have been previously applied by regulatory bodies in Member States.

Several types of monitoring will be introduced and consolidated as follows:

1   Surveillance monitoring

Surveillance monitoring will supplement and assist in validating risk assessments. It will also be capable of highlighting any distinctive trends in the overall quality of the environment. Such changes must be appraised in the light of any coincidental land use changes, climate change and linked human impact changes.

2   Operational monitoring

This type of monitoring programme is focused on those water bodies that are most at risk of failing to meet the directive's considered environmental objectives. For surface waters, this may require attention to diffuse and point source pollution, with 'consent to discharge' conditions being part of the background criteria. These intensive-style programmes may utilise data modelling as part of their outputs, and assist in the identification of remedial measures to improve status.

3   Surface water investigative monitoring

The need for investigative monitoring may emerge from scrutiny of the data produced by surveillance and operational monitoring programmes. Obviously, operational monitoring is devised to use indicators sensitive to known influences. Consequently, investigative monitoring is applied to establish the cause and effect of trends and sudden changes in conditions and environmental quality, and also to help devise short-term management control measures. Investigative monitoring is always used to assess the effects of pollution incidents, either transient or otherwise, and to attempt to establish cause.

4   Protected area monitoring

There are two distinctive aspects to be considered:

(a) Those water bodies that are designated as drinking water protected areas, and which provide in excess of 100 cubic metres of water on an ongoing daily basis, must be monitored to detect any changes from the normal characteristics and to detect any priority substances or any other substances likely to exert an adverse impact on drinking water quality and public health.

Technical Annex X should list priority substances but, as an interim measure, a draft list of 32 hazardous substances has been prepared. These replace substances previously included on an earlier directive's Black List.

The directive also specifies minimum monitoring frequencies, so that all water bodies serving populations of greater than 30 000 must be monitored monthly prior to the water being treated. The Drinking Water Directive prescribes minimum monitoring frequencies for treated water (abstracted from such water bodies) that is supplied to the general public.

(b) Water bodies designated for conservation purposes, to protect habitats and certain species, will require monitoring programmes. In England and Wales, monitoring is mainly in place due to initiatives by English Nature and other similar conservation bodies, who already regularly assess the condition and prevalence of habitats and species included in the Natura 2000 list of Special Areas of Conservation. Freshwater, estuarine and coastal waters are included in this list.

5   Groundwater monitoring

The directive requires the development of a programme that will provide reliable data on both the quality and quantity of all groundwater aquifer systems. Some modelling will be required to establish the relationship between rates of replenishment, recharge and abstraction, or general natural loss to riverine systems and terrestrial ecosystems.

In the past, certainly in the United Kingdom, attention has been focused on those aquifer systems that are used as resources for drinking water supplies and industrial usage. However, the directive also requires groundwaters that may support surface water ecosystems to be identified. The multiplicity of numerous individual small private water supplies also need to be quantified and listed – an onerous task!

6   Monitoring general aspects

The design, frequency and range of parameters included in monitoring programmes must reflect the information required to assist in promoting and achieving suitable environmental objectives. Reliability of the techniques used in the monitoring sequences of sampling, observations, analysis and data handling is also a vital component. The directive requires levels of confidence for all data included in River Basin Management Plans to be reported. Obviously, the criteria used to interpret the results of monitoring programmes must be carefully selected and applied in order to obtain consistency of approval throughout and between Member States.

SAQ 17

Describe the purpose of each type of monitoring programme in the Water Framework Directive, to support the evolution and review of River Basin Management Plans.

2.7 General aspects

Article 9 of the directive requires the costs of water services to be recovered, with the pricing policies to include adequate incentives for users to utilise water resources more efficiently. Existing charging regimes can continue provided they do not compromise the environmental objectives of the directive. In all situations the 'polluter pays principle' needs to be acknowledged.

Currently many charges to users are not volume based and may be only a minor part of their operational expenditure, so dramatic changes to charging regimes could be seriously damaging to economic stability in some regions of Member States.

Given the significant implications of the Water Framework Directive in achieving an improved and more balanced environment, it is worth noting that the directive describes pollution as:

> ... the direct or indirect introduction, as a result of human activity, of substances or heat into the air, water or land which:
>
> - may be harmful to human health or the quality of aquatic ecosystems, or terrestrial ecosystems directly depending on aquatic ecosystems;
>
> - results in damage to material property;
>
> - impairs or interferes with amenities and other legitimate uses of the environment.

Article 18 commits the European Commission to publishing a report on the impact of the implementation of this directive within 12 years initially, and thereafter every six years. The details to be included in the report are outlined in Clause I of Article 18. It also requires a conference to be convened to discuss progress, with the leading participants to include representatives from European organisations and academic institutions, technical experts, social, economic and consumer bodies. The proceedings of this conference are to be published in the interests of transparency and to disseminate knowledge to all concerned.

3   PRETREATMENT

> **READ**
>
> Set Book Chapter 4: Preliminary treatment

This part of water treatment is as important as the similar processes described in T210/T237 *Environmental Control and Public Health* for sewage treatment and, as can be seen, there is a strong resemblance between the techniques used. In sewage treatment, coarse and fine screens followed by grit removal are used to protect pumps and piping and to reduce the load on the remainder of the purification process.

In raw water treatment, coarse intake screens are used, which have similar diameter and spacing to those used in sewage treatment. The fine screens used in water treatment have approximately half the spacing of those used for sewage treatment. The efficiency of fine screens or microstrainers are affected by the fineness of the fabric used and the quantity of solids in the raw water. There are, in all, six factors involved in the flow capacity of an automated strainer:

1   rate of flow of water;

2   loss of head or pressure difference across straining medium;

3   porosity of medium;

4   effective submerged area of straining medium;

5   speed of drum, disc or band;

6   condition of fluid with regard to suspended solids.

The major civil works, complexities and expense associated with fine screens to treat large flows has prompted new techniques to be developed. The wedge wire screen is very simple and is easily cleaned. It requires much less in the way of civil works for installation than an equivalent drum screen.

In water treatment plants, it is very common to have an untreated water storage reservoir holding upwards of seven days' total capacity before treatment. These reservoirs can also be used to even out variations in river flow and help to tide over periods of diminished or suspended river abstraction, e.g. during drought or flood conditions, or during a pollution incident.

The advantages of water storage are:

1   a smoothing out of fluctuations in water quality;

2   a reduction of pathogenic bacteria and oocysts of *Cryptosporidium* and *Giardia*, if present;

3   a reduction of suspended solids;

4   a reduction in colour, ammonia and pesticide levels;

5   oxidation of organic compounds.

The abstraction of water is regulated in order to avoid detrimental effects on the source river. Thus minimum flow in the river downstream of the intake point is specified.

There are, however, disadvantages to consider:

1   There is a tendency for the reservoir to silt.

2   Gulls are known frequenters of sewer outfalls, sewage works, and waste
    tips. At dusk they are known to land on water storage reservoirs. Apart
    from their own faecal pollution (50–100 $\times$ 10^6 *Escherichia coli* per gull
    per day) there is also a danger of specific infective pollution from their legs
    and plumage. In a survey of waters in Lancashire and Cheshire it was
    found that one impounding reservoir was polluted by gulls to the extent
    that 5 out of 6 samples analysed contained *Salmonella*. In the Strathclyde
    region of Scotland the problem has been alleviated by broadcasting
    recordings of the distress calls of gulls to disperse them.

3   In deep reservoirs, stratification can occur, with dissolved iron and
    manganese levels being high in the hypolimnion. Drawing water from the
    epilimnion overcomes any problem. This requires arrangements to draw off
    water from different levels. Low levels of water in the reservoir due to
    excessive demand can obviate the advantages of this arrangement during
    summer periods.

4   There is a possibility of algal blooms occurring. Rivers in the south of
    England are particularly prone to algal blooms as they are generally slow
    flowing, alkaline and contain a considerable concentration of calcium
    bicarbonate. This, along with the nitrates and phosphates present, leads to
    eutrophic conditions. Two-thirds of the water supply of England and Wales
    and over 95% of the water used in Scotland is taken from surface waters in
    which algae can grow! The flora in rivers and streams are different from
    those existing in standing water. So one approach is to store the water long
    enough for the river algae to disappear but not long enough for other
    species to develop. The average storage time of about 10 days is suitable
    for this type of control. (You may be interested to know that research is
    being carried out here in the Open University on the use of rotting barley
    straw to inhibit the growth of algae in freshwater – a case of
    biomanipulation! The technique is widely used by local authorities and
    water companies to control algal growth in, say, water fountains.)

The EU, through the Urban Waste Water Treatment Directive, has addressed the
issue of eutrophication in inland waters, by imposing limits on nitrates and
phosphates discharged from sewage works to sensitive waters (i.e. waters prone
to algal blooms).

Intermittent destratification has been found to reduce algal levels in a reservoir.
Destratification can be achieved by pumping air into the lower region of a
reservoir such that an upward current is induced.

The chemical method of controlling algal growth by dosing with copper
sulphate is not used, due to the toxicity of the chemical to fish.

In some water treatment plants, ozone is used to kill algae before being
removed by sedimentation, filtration or flotation.

In the past, chlorine dosing was carried out early in the treatment process to
eliminate algae, but this is rare now, due to the formation of THMs which it
facilitates. Chlorine can, however, be used to treat groundwaters which are low
in THM precursors to oxidise iron, manganese or ammonia.

An alternative to oxidation with chlorine is aeration. This can also reduce high
levels of carbon dioxide or any hydrogen sulphide present (as can be the case
with groundwaters). Various types of aerator are described in Chapter 4 of the
Set Book.

A combination of a rapid gravity sand filter and microstrainer is sometimes used for algae removal. The former traps the larger species, leaving the remainder to be removed by the microstrainer.

SAQ 18

Referring to Table 4.1 of the Set Book, suggest how the levels of colour, turbidity, ammoniacal-nitrogen, BOD and coliforms might have been reduced through storage of water.

SAQ 19

Referring back to T210/T237 *Environmental Control and Public Health*, suggest how screenings from water treatment plants might be disposed of.

4 COAGULATION AND FLOCCULATION

4.1 Introduction

> **READ**
>
> Set Book Chapter 5: Coagulation and flocculation

Raw water storage and microstraining will have largely removed the suspended matter but the water may still be turbid and coloured. This can be caused by the presence of colloidal particles which include clays, starches, proteins, metal oxides and bacteria, all of which tend to carry a negative charge. Thus most colloidal particles tend to repel each other. This, together with interactions between the colloidal particles and the solvent water molecules, prevents the colloids from aggregating and settling. If the colloidal materials can be artificially induced to adhere to each other or coagulate, they can be removed by sedimentation, and this is often more economical than using filtration. The required aggregation is achieved by first adding the desired amount of coagulant and mixing rapidly to give the first reaction of coagulation. After mixing, the water is stirred gently and the initially coagulated particles grow in size as further particles adhere to form flocs. This is the process of flocculation.

Colloidal systems can be destabilised or coagulated in four different ways:

1   compressing the double layer surrounding the colloidal particle by adding ions, and thereby reducing the repulsive forces between particles;

2   charge neutralisation, by adding ions of an opposite charge to the one on the colloidal particles. These, when adsorbed by the colloidal particles, result in a reduction in the surface charge on the latter, thereby facilitating agglomeration, but excess dosing can result in charge reversal on the colloidal particles and thus their redispersal;

3   entrapment in a precipitate. Aluminium or iron salts form hydroxide flocs using colloidal particles as nuclei. The hydroxide floc can then entrap other colloidal particles, especially when moved around in the flocculation state;

4   particle bridging. In this process, large organic molecules with multiple electrical charges neutralise particles and hold them. They thus act as a bridge between particles. They are known as polyelectrolytes. The flocs formed by polyelectrolytes can be broken up by excessive agitation. As such, polyelectrolytes are more commonly used as flocculant aids in flocculation tanks.

As mentioned above, flocculation is the bringing together of coagulated particles to form large flocs, which can then be removed by settlement, flotation or filtration.

For particles smaller than 0.5 µm, Brownian motion (small movements of the colloidal particles due to bombardment by water molecules) assists in flocculation. Once particles are bigger, flocculation has to be facilitated by slowly stirring the water. The power input of this process has to be carefully controlled – too much will break up the floc, and too little will not have the desired effect.

For coagulation using metal salts, rapid mixing is required. This is because the mechanisms of double layer compression and charge neutralisation are most effective when the salts are present as ion complexes. The salts are in this form for only a second or so, and thus rapid dispersal is required before insoluble salts are formed. For coagulation using bridging or entrapment, the time span is longer and hence rapid mixing is not critical. Flocculation requires slow mixing, so that flocs are not broken up.

The intensity of mixing can be represented by the velocity gradient G. Velocity gradients are created when stirring creates differences in velocity between compartments in the water being treated.

Where mechanical agitation is used, the velocity gradient G can be expressed by the equation

$$G = \left(\frac{P}{\eta V} \right)^{\frac{1}{2}}$$

where

   P is the useful power (watts) consumed in the tank

   V is the volume of fluid (m^3)

   η [pronounced 'eta'] is the dynamic viscosity (N s m^{-2} or Pa s)

The initial mixing can be achieved by using high-speed impellers or pump-induced turbulence. If mechanical means are not used, mixing can be achieved by using sudden flow reversal against a baffle wall (as in a static mixer, shown in the Set Book as Figure 5.3) or by free fall over a weir. All of these methods can provide adequate mixing.

The higher the G value, the better the mixing. For minimal use of metal salt coagulant, a G value of over 5000 s^{-1} is required. For polyelectrolytes, values of 400–1000 s^{-1} are adequate.

Flocculation will take place if the water is allowed to pass along a sinuous inlet channel at a velocity of about 0.3 m s^{-1}. There should be continuous changes in direction, each change requiring the water flow to alter by 180°. There should be a smooth flow, as turbulence or change in velocity could break the floc. Mechanical flocculation is encouraged by vertically oscillating or horizontally revolving paddles in a flocculating chamber.

In the theory of flocculation, the rate at which it takes place is directly proportional to the velocity gradient G. If the retention time (total tank volume divided by flowrate) in the flocculation chamber is T seconds, then the extent of flocculation which takes place, or the number of particle collisions which can occur, will be given by the dimensionless expression GT (known as the Camp number).

For flocculation, G values of 20–75 s^{-1} and contact times of 10–60 minutes are required.

EXAMPLE I

A flocculator treats a flow of 80 litres per second and has been constructed with the following dimensions: depth 3 m, width 4 m, length 18 m. The power required for the motor is 700 W of which 490 W is used to move the paddle. The dynamic viscosity of the water, which has an average temperature of 20 °C, can be taken as 1.01×10^{-3} N s m^{-2}.

Calculate:

(a) the mean velocity gradient G

(b) the Camp number (GT).

ANSWER

(a) Note that the equations for G and T are included in the Water models on the Resources DVD.

$$G = \left(\frac{P}{\eta V}\right)^{\frac{1}{2}}$$

where

P is actual power consumed in moving the paddle $= 490$ W

$\eta = 1.01 \times 10^{-3}$ N s m^{-2}

$V = 3 \times 4 \times 18 = 216$ m^3

Thus

$$G = \left(490/1.01 \times 10^{-3} \times 216\right)^{\frac{1}{2}}$$

$$= \left(490/0.218\right)^{\frac{1}{2}}$$

$$= \left(2247.7\right)^{\frac{1}{2}}$$

$$= 47.4 \text{ s}^{-1}$$

(b) Retention time T is total tank volume/flowrate

Flow $= 80/1000$ m^3 s^{-1}

$= 0.08$ m^3 s^{-1}

Thus $T = 216/0.08$ s

$= 2700$ s

Thus $GT = 2700 \times 47.4$

$= 1.28 \times 10^5$

As explained in the Set Book, the useful power input for sinuous channels can be expressed by the equation

$$P = Q\rho gh$$

where

Q is the rate of flow (m^3 s^{-1})

ρg is the weight of water per unit volume (N m^{-3}) (ρ is pronounced 'rho')

h is the head loss (m)

For channels with baffles the extra head loss h in the channel is found from

$$h = \frac{n v_1^2 + (n-1) v_2^2}{2g}$$

where

$(n-1)$ is the number of baffles

v_1 is the velocity between baffles (m s^{-1})

v_2 is the velocity at baffles slot (m s^{-1})

g is the acceleration due to gravity (m s^{-2})

The Set Book describes various types of mixers. For rapid mixing the following are commonly used:

■ weirs or flumes

■ paddle or propeller mixers

■ turbine mixers

■ static mixers.

For flocculation, paddle flocculators, sinuous or baffled channels, or sludge blanket flocculators are used.

Finally in Chapter 5, the jar test is described. This is the standard test for determining the optimum coagulant dose and pH for treatment of water containing colloidal matter. The jar test is vitally important in checking the conditions for effective coagulation, especially when the quality of the source water is rapidly fluctuating due to storm conditions or snow melt.

4.2   Coagulants

READ
Set Book Chapter 6: Coagulants and coagulant aids

In Section 4.1, we learnt that there are four methods by which colloids are destabilised. Each of these methods requires certain properties of the coagulant, outlined below.

■ For double layer compression, the coagulant should be capable of generating a significant concentration of cations. Trivalent aluminium and ferric salts are very effective.

■ For charge neutralisation, cations are required. Aluminium and ferric salts can again be used.

■ For particle bridging, while theory suggests that long cationic molecules would be best, in practice both anionic and cationic polyelectrolytes can be used.

■ For enmeshment in a precipitate, compounds forming a hydroxide floc or a carbonate precipitate can be used. Suitable candidates are aluminium and ferric salts, calcium hydroxide and magnesium carbonate.

Aluminium or ferric salts dissolved in water go through a series of reactions before finally precipitating as hydroxide floc. The intermediate products are very effective in double layer compression and in charge neutralisation. This point was raised earlier when the importance of rapid mixing was discussed.

The control of pH is important in coagulation, the aim being to maintain it at the level at which the solubility of ferric or aluminium hydroxide is at its lowest. While this pH value can be derived theoretically (see Figure 6.1 in the Set Book for the solubility of aluminium hydroxide at different pH values), it is usual to determine this by conducting a series of jar tests.

Ferric chloride has a minimum solubility over the pH range 7–10, and the solubility change outside this range is less than for aluminium sulphate. This means that ferric compounds can work with less precise pH control than that required for successful use of aluminium-based compounds.

4.2.1   Ferric sulphate

The popularity of ferric sulphate as a coagulant increased in the UK in the 1990s for several reasons:

- the limit on aluminium in drinking water became more stringent
- there was concern over a possible link between aluminium in water and Alzheimer's disease
- there was disquiet over the Camelford incident, in which a tanker delivering the chemical reagent aluminium sulphate accidentally discharged it into a treated water tank. The contaminated water was consequently distributed to the population of Camelford and the surrounding area.

4.2.2   Polymerised aluminium and iron salts

These are inorganic polymers, such as polyaluminium chloride and polyaluminium silicate sulphate, with very high molecular weights. They have several advantages over aluminium sulphate:

- they are more effective at low temperatures
- they lead to faster floc formation
- they require lower dosage rates
- they offer savings in pH adjustment chemicals.

4.2.3   Polyelectrolytes

Polyelectrolytes are long-chain organic molecules with chemical groups attached along the length of the chain which become charged when the molecule is dissolved in water. These groups can be cationic (positive charge), anionic (negative charge), or non-ionic (zero charge). The positive and negative charges bring the small particles in the water together into large agglomerates.

Polyelectrolytes are now used widely in water and sewage treatment, where they may be used as primary flocculants to replace inorganic coagulants, totally or partially, and thereby reduce the mass and the cost of sludge disposal. They may also be used alongside aluminium and iron coagulants to improve their effectiveness.

They are usually made of synthetic chemicals (e.g. polyacrylamides and polyamines) but there are also some made from natural compounds such as starch. Polyelectrolytes made from polyacrylamide can be made anionic, cationic, or non-ionic, and the strength of the charge and the size of the molecule can be made to specification. Polyelectrolytes are available as powders, beads or liquids. They are normally delivered as a powder. They are made into a low-strength solution (typically 0.5%), to ensure that they can be adequately dispersed.

According to the type and method of manufacture, polyelectrolytes may have molecular weights between 5000 and 20 000 000 and have a varying number of charged sites along the molecule, giving different charge densities. In the solid form the molecule is coiled, and as it dissolves in water the attached groups become charged. The repulsive forces between similar charges cause the molecule to uncoil.

After the addition of a polyelectrolyte solution to a water, the charged sites must be brought into contact with the particles in suspension by vigorous mixing. Once the polyelectrolyte molecule has adsorbed the solid particles, the reaction is virtually irreversible. Vigorous mixing is essential to disperse the polyelectrolyte solution through the flow and prevent local overdosing.

The amount of extension of the polyelectrolyte chain depends on its charge density and is also affected by the pH and the salinity of the water. As the chemical groups become attached to solids, they lose their charges and the molecule tends to contract to its original coil, drawing the attached solids into a coherent floc.

Although polyelectrolytes themselves are not toxic, some of the monomers from which they are made, notably acrylamide, are. The monomer is soluble in water and is not adsorbed by solids, so that in the solid/liquid separation which follows flocculation the residual monomer goes with the released water.

The use of polyelectrolytes in the treatment of water is subject to careful control by the Drinking Water Inspectorate. The following conditions apply for polyacrylamide or acrylamide/acrylate copolymers:

1   No batch must contain more than 0.025% of free acrylamide monomer based on the active polymer content.

2   The dose used must contain on average no more than 0.25 mg l^{-1} active agent.

3   An upper limit for the content of the free acrylamide monomer must be stated by the supplier for every batch.

4   The method of analysis for the free acrylamide monomer must be that developed by the Laboratory of the Government Chemist.

When used as a primary flocculant, the main mechanism by which the polyelectrolyte functions is that of charge neutralisation. Cationic polyelectrolytes are much more expensive per kilogram active agent than aluminium or iron coagulants, but because much lower weight dosages are required, they can be cheaper to use. The sludge volumes obtained by using polyelectrolytes are smaller and so the costs of sludge treatment and disposal are less and there is the added advantage that smaller make-up tanks and dosing pumps are required.

Highly coloured waters require a polyelectrolyte dose roughly proportional to the colour, whereas by using aluminium sulphate the demand rises less proportionally. Therefore, for highly coloured waters, e.g. from peaty soil, the most economical practice is often to use a combination of polyelectrolyte and aluminium or iron salts. Because the required dose of coagulant is dependent on the quality of the raw water, it is important to check the dose required in the laboratory on a regular basis. This emphasises the usefulness of regular jar tests.

Where polyelectrolytes are used as flocculant aids, the charge and the charge density are not so important and thus the polyelectrolytes used tend to be of much higher molecular weight and lower charge density than those used as primary flocculants. As a general rule, the higher molecular weight products function better, although this must be balanced against cost and the potential handling problems caused by the greater viscosity of solutions of high molecular weight material.

Table 6.2 in the Set Book summarises the performance of alum, ferric salts, and polymers (polyelectrolytes) in treating four different types of raw waters.

4.2.4   Weighters

If enmeshment is chosen as the process whereby the colloidal particles in a given water are removed, a very low concentration of particles will require a high dosage of coagulant. Adding fine materials, such as bentonite or fuller's earth (these are types of clay) to increase the solids concentration may result in less coagulant being needed to clarify the water. These materials are called weighters. They can also be used where polyelectrolytes are utilised as the main coagulant, to help weight the flocs down in order to speed up their settlement.

4.2.5   Control of coagulation and flocculation

Coagulation and flocculation processes need to be controlled to ensure optimal treatment of the raw water, at minimal cost. This calls for strict dosage control of added chemicals. The traditional method of determining dosage was through jar tests but this can be time consuming. Automated methods of determining dosage have been devised. One system that is available is the streaming current detector (SCD). The instrument takes a sample of coagulated water and deduces the charge on the particles in it, by measuring the current the sample generates when the particles go past a reciprocating piston. The dose of coagulant can then be controlled to ensure coagulation of the particles. While attractive in theory, the SCD system has practical difficulties, and as such is not widely used in the UK.

Another approach is to establish relationships between historical and current data on raw water quality, coagulant dose and pH value, and treated water quality. The correct dose for any given combination of parameters is then ascertained, and this can be dispensed, either manually or automatically.

Yet another option is to measure a relevant water quality parameter (e.g. turbidity, pH, colour, residual coagulant, etc.) downstream of the coagulation process, and make the necessary adjustments to the dosage of the coagulant. This can be achieved automatically by using closed-loop control.

The storage of coagulants has to be in tanks which will withstand their specific requirements. All tanks and associated transfer pumps need to be bunded to contain any spillages or leaks. Table 6.3 in the Set Book gives details of storage and dosage requirements of the major chemicals used in coagulation and flocculation.

4.2.6   The Sirofloc™ process

This is a patented process (Figure 6), developed in Australia, which uses finely divided magnetite (Fe_3O_4, a naturally occurring ore of iron) to adsorb colloidal matter (including colour-conferring compounds) and some metal species (Fe and Al) from water. Magnetite, which can be readily magnetised and demagnetised, exhibits a charged surface when in aqueous suspension. The charge is positive at low pH and negative at high pH, with the zero charge (or isoelectric point) being between pH 5 and pH 7.

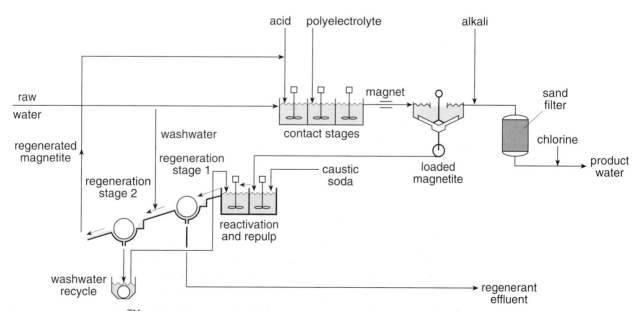

Figure 6   The Sirofloc™ process [Courtesy of WRc Swindon Plc]

In the Sirofloc^{TM} process, acid and magnetite are mixed with the raw water in contact tanks where the magnetite adsorbs colour, negatively charged colloidal particles and metals. A small amount of polyelectrolyte may also be dosed to enhance adsorption of colour and turbidity. The magnetite, having a diameter of only 1–10 μm, does not settle readily by gravity. When passed through a magnetic field, however, aggregates are formed and these settle rapidly. The magnetite is recovered from the settlement tank and regenerated by raising the pH using caustic soda, which causes desorption of the impurities. The magnetite is then washed, using rotating magnetic drum separators, before being reused. The alkaline waste containing the raw water impurities is disposed of separately.

The product water from the Sirofloc process has to be filtered to remove any carry-over of magnetite.

SAQ 20

If a raw water was highly turbid with a low alkalinity, which coagulant might be best?

SAQ 21

Calculate the flowrate of 8% Al_2O_3 solution required to dose a flow of 1000 m^3 d^{-1} with 1 mg l^{-1} Al.

SAQ 22

Polyelectrolytes can be used as primary flocculants or as flocculant aids. What differences are there in the polyelectrolytes used for each purpose?

SAQ 23

In SAQ 21, the amount of colloidal matter removed was 7 mg l^{-1}. Calculate the volume of sludge produced per day, assuming a sludge solids content of 2.5%, and a sludge density of 1100 kg m^{-3}.

SAQ 24

A flocculator treating 25×10^6 litres per day has a total volume of 650 m^3 and an average velocity gradient G of 25 s^{-1}. If the viscosity of the water is 1.15×10^{-3} N s m^{-2}, calculate:

(a) the retention time in the flocculator;

(b) the Camp number GT;

(c) the electrical power required.

SAQ 25

Water is flocculated as it passes along a sinuous channel with round-the-end cross walls. The water velocity between the cross walls is 0.3 m s^{-1} and 0.45 m s^{-1} round the end of the cross walls. The power input is found to be 1 kW and the rate of flow 0.53 m^3 s^{-1}. As the water has a temperature of 10 °C, the weight of the water can be taken as 9.80×10^3 N m^{-3}.

Calculate the head loss achieved through the channel and the number of baffles likely to be present.

---

5   THEORY, PRINCIPLES AND METHODS OF CLARIFICATION

5.1   Introduction

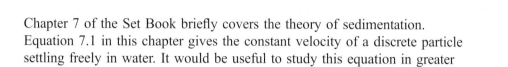

READ

Set Book Chapter 7: Theory and principles of clarification

Chapter 7 of the Set Book briefly covers the theory of sedimentation. Equation 7.1 in this chapter gives the constant velocity of a discrete particle settling freely in water. It would be useful to study this equation in greater depth as it is relevant to practical applications in both the Water and Air blocks.

Let us start by imagining the release of an object such as a football or a feather in air. Both will fall to the ground, although the feather will fall more slowly because it is influenced to a greater extent by air resistance.

Any solid object moving through a liquid or a gas at a speed v experiences a force in the direction opposite to its motion. This is the drag force which, except at very low speed, is the net effect of the speed, size and shape of the object as well as the density, viscosity and compressibility of the fluid. The resistance experienced by a car at high speed, an aeroplane or a falling raindrop are examples of air drag. This resistance is roughly proportional to the square of the object's speed v; to the density ρ of the fluid; and to the area A of the object in a plane perpendicular to the object's motion. The drag force F_D may be written

$$F_D = \tfrac{1}{2} C_D \, A \, \rho v^2 \tag{5.1}$$

where C_D is the empirical drag coefficient, which is a dimensionless quantity.

You will notice that the above equation does not include a viscosity term, but we know, for example, that honey is more viscous than water. Viscosity can be defined as the resistance of a fluid to shear forces and hence to flow. This is referred to as absolute or dynamic viscosity. A small stone dropped into a glass of water will fall more quickly than if dropped into a jar of honey. In fact, for a small particle moving slowly through a fluid, Equation 5.1 is not valid. In this case, the drag force is proportional to the viscosity of the fluid and also depends linearly on the velocity. For a sphere of radius r falling in a fluid of dynamic viscosity η the equation becomes

$$F_D = 6\pi r \eta v \tag{5.2}$$

The question now to be answered is how small a particle must be, or how slowly it must move in order for Equation 5.2 rather than Equation 5.1 to be valid. The answer is found to depend on a dimensionless parameter known as the Reynolds number, named after Osborne Reynolds (1842–1912), a scientist and mathematician who investigated the effect of viscosity in pipe flow. Reynolds number (R_e) is defined as

$$R_e = \frac{\rho v d}{\eta} \tag{5.3}$$

where ρ is the density of the fluid, η its dynamic viscosity, d the diameter of the object, and v its speed through the fluid. Different values of the Reynolds number determine different types of flow in which different laws apply to the drag force. Table 14 summarises the situation.

Table 14

Reynolds number	Drag force
0 to 10	Equation 5.2 applies
10 to 300	Transition region
3×10^2 to 3×10^5	Equation 5.1 applies. Drag coefficient nearly constant
$\geqslant 3 \times 10^5$	Drag coefficient varies

In the Set Book, a graph representing the variation of drag coefficient with Reynolds number will be found as Figure 7.1.

Now consider a particle of mass m falling in a fluid. The particle is acted on by the upward drag force and upthrust of the fluid, while the weight of the object acts downwards. The impelling force downwards F is given by

$$F = mg = (\rho_1 - \rho)gV \tag{5.4}$$

where g is the acceleration due to gravity, V is the volume of the particle and ρ_1 and ρ are the densities of the particle and the fluid respectively. The drag force is given by Equation 5.1. The downward component of the force on the particle is now given by

$$F - F_D = mg - \tfrac{1}{2}C_D A\rho v^2 \tag{5.5}$$

If we analyse this relationship qualitatively, after the initial release of the particle, the velocity is low and so the downward force is approximately mg. However, the mass accelerates, i.e. the velocity increases with elapsed time. As this happens, the magnitude of the drag force becomes more significant because it is related to v^2. This will have the effect of slowing down the acceleration until eventually a point is reached where the drag force equals the value of mg. No further acceleration takes place – the particle has reached its *terminal velocity*. Thus from Equation 5.5,

$$F - F_D = 0$$

Hence,

$$mg = \tfrac{1}{2}C_D A\rho v^2$$

or

$$(\rho_1 - \rho)gV = \tfrac{1}{2}C_D A\rho v^2 \text{ (from Equation 5.4)}$$

Rearranging gives

$$v = \left[\frac{2g}{C_D} \frac{(\rho_1 - \rho)}{\rho} \frac{V}{A} \right]^{\frac{1}{2}} \tag{5.6}$$

Substituting the expressions for $V\,(=\,4\pi r^3/3)$ and $A\,(=\,\pi r^2)$ we get

$$v = \left[\frac{8g(\rho_1 - \rho)r}{3C_D\rho}\right]^{\frac{1}{2}}$$

$$= \left[\frac{4g(\rho_1 - \rho)D}{3C_D\rho}\right]^{\frac{1}{2}} \qquad (5.7)$$

where D is the particle diameter.

This is the same as Equation 7.1 in the Set Book.

For low Reynolds numbers,

$$C_D = \frac{24}{R_e}$$

Substituting this into Equation 5.7 gives

$$v = \frac{g}{18}(\rho_1 - \rho)\frac{D^2}{\eta}$$

This is Equation 7.3 in the Set Book, which is Stokes' law.

Using the expression for kinematic viscosity v (pronounced 'new'),

$$v = \frac{\eta}{\rho}$$

the equation becomes

$$v = \frac{g}{18}\frac{(\rho_1 - \rho)}{\rho}\frac{D^2}{v}$$

$$= \frac{g}{18}(s-1)\frac{D^2}{v}$$

where

$$s = \frac{\rho_1}{\rho} = \text{ specific gravity of the particle}$$

The effect of particle size on settling velocity v_s can be shown if we consider two particles with diameters 0.01 mm and 0.06 mm and specific gravity 2.5 settling in water with a kinematic viscosity of 1.010×10^{-6} m^2 s^{-1}.

For the 0.01 mm particle,

$$v_s = \frac{9.81}{18}(2.5-1)\frac{\left(1 \times 10^{-5}\right)^2}{\left(1.01 \times 10^{-6}\right)}$$

$$= 8.1 \times 10^{-5} \text{ m s}^{-1}$$

$$= 0.08 \text{ mm s}^{-1}$$

Carrying out the same calculation for the 0.06 mm diameter particle gives a settling velocity of 2.91×10^{-3} m s^{-1} or 2.91 mm s^{-1}.

For horizontal flow tanks, the settling velocity v_s in m s^{-1} is equal to Q/A where Q is the flowrate (m^3 s^{-1}) and A is the area of the tank (m^2). The derivation of this is given in the Set Book. The value of Q/A is the overflow rate or the surface loading rate. The relationship shows that the efficiency of sedimentation is independent of the depth of the tank. Thus all particles with values greater than Q/A will reach the bottom of the tank before the outlet, and particles with a velocity less than Q/A will be removed from suspension in the same proportion that the velocity bears to Q/A.

It is possible to make use of this equation and Stokes' law to determine the diameter of particles settling out.

EXAMPLE 2

A flow of 25 l s^{-1} of discrete particles of varying size but all with specific gravity 2.5 is allowed to settle in an ideal settlement tank which is 5 m long and 2 m wide. If the kinematic viscosity of the water is 1×10^{-6} m^2 s^{-1}, what will be the diameter of the particles which completely settle out?

ANSWER

$$v_s = \frac{Q}{A}$$

$$= \frac{Q}{L \times W}$$

$$= \frac{25 \times 10^{-3}}{5 \times 2}$$

$$= 2.5 \times 10^{-3} \text{ m s}^{-1}$$

From Stokes' law,

$$v_s = \frac{g}{18}(s-1)\frac{D^2}{v}$$

$$D^2 = \left[\frac{18 v v_s}{g(s-1)}\right]$$

$$D = \left[\frac{18 v v_s}{g(s-1)}\right]^{\frac{1}{2}}$$

$$= \left[\frac{(18)(1 \times 10^{-6})(2.5 \times 10^{-3})}{9.81(2.5-1)}\right]^{\frac{1}{2}}$$

$$= 5.53 \times 10^{-5} \text{ m}$$

$$= 0.0553 \text{ mm}$$

The theory dealt with so far has considered ideal conditions and discrete particles. In water treatment, we are interested in flocculant particles which will aggregate and in so doing alter their settling velocity. For these particles the tank must be sufficiently deep to allow them to settle out. In other words, the settlement is *not* independent of depth. In very shallow tanks there could be interference of settlement by the turbulence caused by excessive horizontal flow.

Settlement in upward-flow tanks is explained in the Set Book. The main point to note is that the settling velocity appears to be dependent only on area but, as in horizontal flow settlement, depth is important. You should pay particular attention to the merits of having adequate depth in both upward- and horizontal-flow basins. Read the sections on shallow-depth sedimentation and dissolved air flotation. The Sirofloc^{TM} process, which you met in Section 3, is also described.

SAQ 26

In Example 2 above for the flow of 25 l s^{-1} it was calculated that particles with a diameter of 0.055 mm would fully settle out. If the same conditions are maintained in the tank, calculate the proportion of particles which have a specific gravity of 2.7 and a diameter of 0.035 mm which will settle out.

SAQ 27

Calculate the theoretical surface area of a horizontal flow sedimentation tank which will remove particles with a settling velocity of 1.8 mm s^{-1} from a flow of 350 l s^{-1} under ideal conditions. In the final design for such a tank, what is likely to be the surface area?

5.2   Settling basins – types and operation

> **READ**
>
> Set Book Chapter 8: Types of clarifiers

Chapter 8 of the Set Book describes a selection of settling basins. As would be expected, not all of the various types of basin described have world-wide popularity. In the UK there would appear to be a preference for vertical upward-flow systems, which have also found favour in the Netherlands and Germany. Horizontal flow systems have equal popularity with vertical flow systems in Belgium and France.

The Set Book describes a wide variety of settling basins but you need only pay particular attention to the following basic designs.

5.2.1   Horizontal-flow sedimentation tanks

This type of basin can be either rectangular or radial and there is no fundamental difference in the hydraulic design of the tanks. The type selected will depend on the personal choice of the designer or on local conditions, e.g. area of land available.

There has been considerable interest shown in the use of inclined plates and tube clarifiers to increase the throughput rates and reduce the size of treatment works. France, Belgium and the Netherlands have also shown particular interest in these innovations. Tube clarifiers and inclined plates can be used to uprate existing tanks.

5.2.2 Upward-flow basins

The popularity of this type of design is due to the following reasons:

1   In many waters the settleability of sludge improves with age. Retained and recycled sludges within a clarifier form good nuclei for the growth of new flocs and increase settling velocity.

2   The high concentration of particles of various sizes provides relative movement and collision opportunities to improve flocculation.

3   The density difference between the slurry zone (also known as the floc blanket) and the clean water above it enhances the hydraulic stability of the flow through the clean water or final settling zone.

4   The amount of sludge (or floc blanket) sliding down the sloping sides of the hopper into the incoming water is sufficient to form a mixture having the same concentration and density as that of the slurry (or floc blanket) already in the slurry zone, thus avoiding unstable density conditions.

Upward-flow basins all have the common feature of the raw water and the coagulants being filtered as they pass through a layer of previously deposited or settled sludge or floc blanket. The design of the Accentrifloc™ clarifier (Figure 8.5 of the Set Book) incorporates many of the design features required for this type of basin. You should note that this type of tank has a flat bottom, and this overcomes the weakness of hopper-bottomed basins, which need to penetrate deep into the ground. The sludge or floc blanket can take several days to establish, and in each case the sludge is recirculated and the excess sludge produced bled off from the system.

The clarifying action which takes place in upward-flow tanks is dependent on the existence of a uniform and thick layer of sludge or floc blanket. The constant flow through such tanks makes this difficult to achieve, but the problem has been overcome by the use of the Pulsator™ tank which allows the raw water and coagulants to enter the tank in controlled pulses, giving only a small displacement of the treated water. (You will see more details of the Pulsator in the T308 Resources DVD relating to the project.)

5.3 Practical considerations and choice

In reading Chapter 8 of the Set Book, you should pay particular attention to the fact that generous safety margins have to be incorporated into even the most detailed theoretical design. In addition to the information given in the Set Book, the retention times and surface loading rates for horizontal tanks can be up to three times greater than those indicated in still water, while for upward-flow basins the upward velocity should be half that of the settling particles. You should also specifically note the difficult conditions which may have to be overcome in water treatment.

5.4 Air flotation

In air flotation, air comes out of solution and becomes attached to floc particles, allowing them to float to the surface, where they form a scum which can be scraped off. The size of the air bubbles is microscopic; a mist is thus formed. This is an important aspect, as the air comes out of solution. The advantages of this type of treatment are as follows.

■   Flotation units are usually smaller than normal clarifiers.

■   Retention times are reduced.

■ The scum produced contains less water than settled sludge and thus requires less dewatering prior to disposal.

■ It is particularly good at removing algae and other low-density solids.

■ It can be introduced into existing sedimentation systems.

The disadvantages of air flotation are as follows.

(a) It requires air and water pumps which have a constant energy demand.

(b) It requires regular maintenance.

(c) The considerable extra operating costs of (a) and (b) have to be set against reduced capital costs.

5.5   Selection of settling basin (clarifier) type

The most suitable type of clarifier will depend on the characteristics of the water and the variability of the pattern of the water demand.

The performance claimed for proprietary designs can be achieved but will require good conditions, expert control (competent engineers and process chemists) and careful use of chemicals. Therefore, it may be inadvisable to attempt the use of sophisticated water treatment processes if these conditions cannot be met.

The appropriate clarifier will be the one which meets the performance requirement at the lowest cost. The following are general 'rules' which can be used as guidelines.

1   Waters with low alkalinity, low turbidity and high colour often produce a light, slow-settling floc, especially in cold climates. The floc settling velocity can be increased if the floc is aged for several days. Thus a system of sludge reclamation and recycling can be of value.

2   Turbid water usually forms a good floc which will settle rapidly because of the influence of the clay particles present. For this type of water a wider choice of systems exists. Horizontal and shallow-depth tanks perform well in stop–start conditions, while tanks using sludge- or floc-blanket clarifiers operate best when the flow is almost constant. Stop–start conditions are best avoided, as are rapid fluctuations in flowrate through the process. Stable flow enhances the effectiveness of the treatment overall.

   If shallow-depth sedimentation tanks are used, there must be good flocculation because the water does not remain in the tank long enough for flocculation of the suspended solids to occur. Shallow-depth sedimentation also requires very good flow control, to prevent turbulence. Wind conditions can also adversely affect uncovered sedimentation tanks.

3   It is very difficult to determine the best system if temperature fluctuations exist. There are, however, two factors which could help in achieving the required conditions for sludge- or floc-blanket clarifiers. First, good mixing in the sludge-blanket zone will assist in maintaining an even temperature in that zone. Secondly, a shallow clean-water zone will limit the potential energy released as the warmer water rises through the ambient cold water, and this will reduce the disturbance caused by convection currents.

4   Direct filtration may be used when the raw water has a low turbidity, or is cold or supersaturated with air and a settleable floc is formed only with difficulty. In this process the water is dosed with coagulants which are allowed to react before it is applied directly to the filters. The savings resulting from the omission of the clarification process should offset the extra cost required for more frequent filter cleaning.

EXAMPLE 3

A water treatment works has a maximum flow of 0.5 m^3 s^{-1}. The works uses a sedimentation tank which has a surface loading rate of 2.5 l s^{-1} m^{-2}. The length-to-width ratio of the tank is 4. Calculate the surface dimensions of the tank.

ANSWER

Surface loading rate $= 2.5$ l s^{-1} m^{-2}

(i.e. each second 1 m^2 treats 2.5 litres)

$$= 2.5 \times 10^{-3} \text{ m}^3 \text{ s}^{-1} \text{ m}^{-2}$$

$$\text{Surface loading rate} = \frac{Q}{A}$$

$$Q = 0.5 \text{ m}^3 \text{ s}^{-1}$$

$$A = \frac{Q}{\text{surface loading rate}}$$

$$= \frac{0.5}{2.5 \times 10^{-3}}$$

$$= 200 \text{ m}^2$$

If width of tank $= w$, length is $4w$, and area is $4w^2$

$$\therefore 4w^2 = 200$$

$$w = 7.1 \text{ m}$$

$$\text{length} = 4w = 28.4 \text{ m}$$

Therefore the tank dimensions would be: width 7.1 m and length 28.4 m. In practice, these would be rounded off, with the width chosen so as to enable readily available sludge-scraping equipment to be used.

SAQ 28

An Accentrifloc^{TM} clarifier is 94.2 m in circumference. The inner circular flocculation chamber is 10 m in diameter. The flow through the tank is 0.6 m^3 s^{-1}. Calculate the surface loading rate at the outer chamber.

SAQ 29

(a) Give two advantages of tube settlers over inclined plate settlers.
(b) What are the advantages of air flotation?

SAQ 30

A clarifier is required for a small water treatment plant. Taking the following factors into account, use Table 8.1 in the Set Book to select a suitable clarifier:

    effectiveness with algae
    effectiveness on small works
    ability to withstand sudden changes in water quality
    performance with unskilled operators
    low maintenance requirements.

SAQ 31

(a) Why is the outlet weir an important factor in settling basin design and how can problems associated with outlet weirs be overcome?
(b) What are the major design features of the Accelator type of clarifier and how does it resemble the system used in the biological oxidation of sewage?

6 FILTRATION

> **READ**
>
> Set Book Chapter 9: Filtration

6.1 Introduction

Chapter 9 considers the final process of water clarification, i.e. filtration. Filtration may be unnecessary for most groundwaters but with the ever-improving standards set for water quality it is essential for surface-derived waters.

The filtration process consists of passing the water through a bed of sand or other media. The suspended matter is retained in the sand and clean water obtained as a product.

It was originally thought that the reaction taking place in a granular filter was simply that of straining. If filters act mainly by straining then we would expect the larger particles of impurities to become lodged in the spaces between the grains, with the smaller openings left trapping the smaller particles. This theory seems to be acceptable until we realise that the surface layer would rapidly become clogged, and at this point very fine material would be forced through the very small openings without being trapped. A granular filter is, however, able to capture very fine material – so mechanical straining can only account for a small part of the filtering action. The main process by which particles are retained is *adsorption*.

Adsorption takes place when particles in the water adhere either to the filter material or to previously adsorbed particles. The method by which particles attach is similar to the coagulation process. If a particle is close to a solid surface then there may be either electrical attraction or repulsion, depending on the surface charges of both the particle and the solid surface. If the water has already been treated to give effective flocculation and sedimentation then it is very likely that the particles will be adsorbed. This process plays a much larger part in the action than simple straining.

If the forces which lead to adsorption are so strong that the impurities are not readily removed during backwashing, the impurities and the filter medium may form mud balls which reduce the efficiency of the filter. Care must therefore be taken to make sure that the forces giving adsorption are weak enough to allow the trapped particles to be released during backwashing.

As particles flow through a filter, there are three ways by which they become attached to the filter material.

1 *Interception* In this process the particle, by chance, comes close enough to the filter surface for adsorption to take place.

2 *Sedimentation* Here the particle is deflected from the path of the water flow by the action of gravity so that it comes into contact with the filter material.

3 *Diffusion* This occurs when the particles are randomly deflected by collisions which are the result of molecular motion.

The effectiveness of interception and sedimentation increases with increase in particle size, while the effectiveness of diffusion increases with decrease in particle size. From this it can be deduced that although a filter can remove large particles by interception and sedimentation, and small particles by diffusion, there is likely to be an intermediate size of particle which will not be easily removed. For most filters this size is about 1 μm.

You should be able to describe the basic types of filter which are discussed in the Set Book.

These are:

1   *Rapid gravity sand filters* These are the most commonly used and involve the utilisation of coagulants. They can treat water with a turbidity of 10–20 NTUs and will operate for short periods with incoming water turbidities of 20–50 NTUs.

2   *Slow sand filters* These work without coagulants and are therefore useful for good quality lake or reservoir water, and so are generally not found alongside settling basins.

6.2   Rapid gravity sand filters

Figure 9.3 of the Set Book outlines the main features of the rapid gravity filter. From your reading, you should be able to explain the construction and action of this type of filter and the different ways in which good control of the flow through such a filter can be achieved.

The sand used for rapid filters should be sharp, hard, clean, siliceous and have a uniformity coefficient between 1.2 and 1.7. The uniformity coefficient is the ratio of the size of sieve which permits the passage of 60% by weight of the sand to that which permits the passage of 10% by weight.

EXAMPLE 4

If a 1.00 mm sieve allowed 60% of a sand to pass and a 0.42 mm sieve allowed 10% to pass, would this sand be acceptable for a rapid sand filter?

ANSWER

The uniformity coefficient would be $1.00/0.42 = 2.4$. so the sand would not be suitable.

There must also be good control over the flow through the filter and the backwash cycle. In normal operation, the inlet valve and the filtered water valve are kept open while all other valves are kept closed. The desired filter rate is usually achieved by fitting a controller to the outlet pipe. The same effect can be obtained by varying the head of water on the filter. These techniques can be applied on rapid gravity filters and also on mixed media filters.

You should be able to state when a rapid filter will require to be washed and how this cleaning process is carried out (see section on 'Rapid sand filters' in Chapter 9 of the Set Book). It is worth noting that the backwashing process can lead to the use, and hence the loss, of a considerable volume of water.

EXAMPLE 5

A rapid gravity sand filter operates at $2.2 \, \mathrm{l \, s^{-1} \, m^{-2}}$ for 23 hours before it requires a backwash. The backwash cycle takes 30 minutes, which includes 15 minutes of actual washing at a rate of $12 \, \mathrm{l \, s^{-1} \, m^{-2}}$. Estimate the percentage of filtered water used in backwashing.

ANSWER

Total volume of water treated in 23 hours $= 2.2 \times 60 \times 60 \times 23 \text{ l m}^{-2}$

$$= 182160 \text{ l m}^{-2}$$

$$= 182.16 \text{ m}^3 \text{ m}^{-2}$$

Backwash water required $= 12 \times 60 \times 15 \text{ l m}^{-2}$

$$= 10.8 \text{ m}^3 \text{ m}^{-2}$$

Percentage water used in backwashing $= \dfrac{10.8}{182.2} \times 100$

$$= 5.93\%$$

The use of multilayer filters and the possible use of a reverse flow filter are explained in Chapter 9 of the Set Book.

6.3 Pressure filters

Pressure filters are described Chapter 9 of the Set Book and you should study Figure 9.8 so that you are aware of the main features of this type of filter and know that they can be used when reasonably clean reservoir water, travelling in a steeply inclined pipeline, requires filtration.

A serious disadvantage with pressure filters is the inability to observe the filtration media and the backwashing process. If there is damage to the underdrains, for example, media might be lost, or the filter may not be efficiently backwashed and this could remain unnoticed for some time until the final water quality becomes seriously affected. There can also be channelling through the depth of the media which will also remain unnoticed.

6.4 Slow sand filters

The main features of slow sand filters are shown in Figure 9.9 of the Set Book. As with rapid gravity sand filters you should be able to describe their construction and mode of operation. You might also have noticed that the mode of action of slow sand filters resembles that of biological filters in sewage treatment (see T210/T237 *Environmental Control and Public Health*). When water is first applied to a slow sand filter there is little or no purification until a biologically active surface mat (*Schmutzdecke*, literally a 'deck of dirt') builds up as a gelatinous mass of bacteria and detritus which breaks down the organic matter. This mat assists in the retention of components causing turbidity and harbours the predators which remove pathogens and excess bacteria.

From the foregoing, it is obvious that slow sand filters may be useful for remote areas, especially in developing countries. Indeed, research has shown that burnt rice husks can be used as a filter medium since the protective coat of rice grains has a silica framework. If some carbonaceous material is left after incomplete burning, it may act as activated carbon and adsorb small quantities of organic material. This is one way of providing a cheap, effective filter and also disposing of unwanted rice husks!

6.5   Other filters

The Set Book describes the 'Dynasand' filter which has an inbuilt constantly cleaning action. It is suitable as a pretreatment filter.

The wound-fibre filter used for removal of *Cryptosporidium* is mentioned, as are diatomite filters used in the field by service personnel and for emergencies.

SAQ 32

A sand was graded and the following results were obtained:

sieve size (mm)	0.30	0.42	0.60	0.84	1.00
% of sand retained	98	90	40	30	20

What is the uniformity coefficient of this sand? Would it be suitable for use in filtration?

SAQ 33

What are the main differences between the operation of slow sand filters and rapid gravity sand filters?

SAQ 34

What are the main construction features of rapid gravity sand filters?

SAQ 35

A treatment works has four rapid gravity sand filters. Each filter is 75 m^2 in area and is backwashed for 6 minutes every 24 hours at a wash rate of 15 mm s^{-1}. During backwashing each filter is out of operation for 30 minutes. What volume of water is required for the backwashing of all the filters during each 24-hour cycle?

---

7   DISINFECTION

READ

Set Book Chapter 12: Disinfection

7.1   Introduction

Disinfection is the killing of pathogenic or disease-causing organisms. The three main classes of pathogenic micro-organisms that are important in water treatment are viruses, bacteria and protozoa. You will have come across the main water-borne diseases carried by such organisms in T210/T237 *Environmental Control and Public Health*. The major disinfectants are chlorine, ozone, UV and chlorine dioxide.

Disinfection is necessary because even effectively filtered water may not be free from bacterial contamination, with the result that disinfection may be required to remove this contamination or reduce it to a negligible amount. Disinfection differs from sterilisation, which implies the destruction of all micro-organisms.

A suitable disinfectant should:

- be able to destroy pathogens at the concentration in which they occur
- be effective in the normal range of environmental conditions, i.e. pH, temperature, etc.
- not be toxic to humans or higher animals, e.g. fish
- provide some residual disinfecting capacity to protect against reinfection of the water in the distribution system (the residual, however, should not give rise to consumer complaint, e.g. taste, odour, etc.)
- be easy to handle during storage and use
- allow simple measurement of its concentration in water (this allows accurate control of addition)
- not be too expensive as large amounts may be required in treatment.

Chlorine is the most widely used disinfectant as it fulfils many of the above criteria. Other disinfectants in use are chlorine dioxide and ozone. Ultraviolet radiation can also be used and has found application in the treatment of small water supplies, and in hospitals where higher than normal water quality is required for specialist use.

7.2   Chlorine as a disinfectant

Compounds of chlorine are believed to kill micro-organisms by rupturing the cell membrane and destroying the enzymes within the cell. The Set Book highlights the fact that there is resistance in some countries over the use of chlorine as a residual disinfectant. If the water is treated such that there is negligible dissolved assimilable organic carbon, and the distribution system is clean and in good condition, a residual may not be required.

Care is taken to reduce the formation of trihalomethanes (THMs). Where residual disinfectants are needed for the distribution system, chlorine, chloramines or chlorine dioxide have been utilised.

The chlorination processes result in either hypochlorous acid or the hypochlorite ion. Hypochlorous acid is a considerably more effective disinfectant than the hypochlorite ion. Together they are called 'free chlorine'. Chloramines can be formed when ammonia is present, and are referred to as 'combined chlorine'. The pH is a critical parameter, affecting both the degree of formation of hypochlorite ions and the type of chloramine formed. The WHO suggests that for effective disinfection the following are needed:

■ a free chlorine residual concentration of more than 0.5 mg l^{-1}

■ a chlorine contact time of at least 30 minutes

■ turbidity of less than 1 NTU, and

■ a pH of no more than 8.0.

Superchlorination is the preferred method of disinfection. This has to be followed by dechlorination to reduce the residual chlorine to about 0.5 mg l^{-1} in order to prevent complaints from consumers concerned about the taste.

The chlorine dose available to kill pathogens can be presented as in Figure 7.

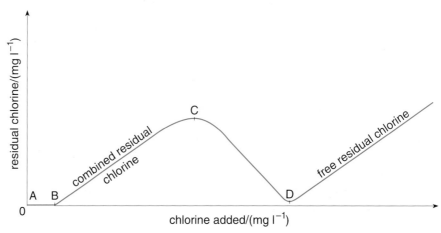

Figure 7   The relationship between quantity of chlorine added to a natural water and the chlorine available for disinfection [Redrawn with permission from the McGraw-Hill Companies, Inc., Metcalf and Eddy Inc, revised by Tchobanoglous (1979)]

The sections A–B, B–C and C–D, and point D, in Figure 7 are defined as follows.

1   A–B. The chlorine reacts with reducing agents. The residual chlorine is low and there will be little disinfection.

2   B–C. The chlorine forms mono-, di-, and trichloramines with ammonia.

$NH_3 + HOCl \rightarrow H_2O + NH_2Cl$ (monochloramine)

$NH_2Cl + HOCl \rightarrow H_2O + NHCl_2$ (dichloramine)

$NHCl_2 + HOCl \rightarrow H_2O + NCl_3$ (trichloramine)

The chloramines are disinfectants but less powerful than chlorine. If they are present in high concentration or have a long contact time, they will destroy pathogens. Monochloramines and dichloramines predominate in most cases and are considered as *combined available chlorine.*

3   C–D. Further addition of chlorine produces N_2O and N_2 (which are not disinfectants) from the oxidation of the chloramines. Possible reactions are:

$NH_2Cl + NHCl_2 + HOCl \rightarrow N_2O + 4HCl$

$$4NH_2Cl + 3Cl_2 + H_2O \rightarrow N_2 + N_2O + 10HCl$$

$$2NH_2Cl + HOCl \rightarrow N_2 + H_2O + 3HCl$$

$$NH_2Cl + NHCl_2 \rightarrow N_2 + 3HCl$$

4 D. At this point all the possible reactions are complete and any subsequent addition of chlorine will appear as *free available chlorine*. Point D is the *breakpoint*. Chlorination is commonly carried out beyond the point D to ensure the presence of free residual chlorine.

If the water contains so much ammonia that breakpoint chlorination would be prohibitively expensive, then the chlorination is only taken to B–C, where a combined chlorine residual is formed.

For waters of satisfactory quality, without excessive colour or turbidity, not subject to wide variations of quality and free from phenolic substances and algae, a chlorine dose of 0.5 mg l^{-1} with a contact time of 30 minutes at pH 6–7 will give the desired disinfection. Chloramines, being weaker disinfectants, require contact times of approximately one hour and can be useful for waters where a chlorine residual has to be maintained for a long time.

When the chlorine is added to a particular water the residual content depends on the dose applied and the contact time. It is necessary to know the conditions to ensure that the pathogens present have been exposed to the appropriate chlorine concentration for the requisite time period. Thus the conditions required for the desired disinfection can be expressed as 'total free chlorine after x minutes shall not be less than y mg l^{-1}'.

Colour, taste and odour problems exist in water supplies and are often caused by the presence of organic molecules, or iron and manganese. Chlorine will oxidise these materials, thus reducing the problem. Care must be taken if phenols are present as these lead to taste and odour problems caused by even low concentrations of chloro-phenolic compounds.

7.2.1   Chlorination before filtration

This technique may be used to reduce the bacterial load in the early treatment of the water, and to oxidise ammonia and the humic/fulvic acids in peaty, highly coloured, surface waters. One adverse effect of this treatment is the production of trihalomethanes (THMs). Chlorine is sometimes added either continuously or intermittently to the water passing through pipes transferring raw water from, say, a river to a reservoir, to deter the colonisation of mussels on the pipe walls. If the mussels become established they can reduce the pipe diameter considerably, causing a reduction in flow. Once established, they are extremely difficult to remove.

Note the information in Chapter 12 of the Set Book which explains how chlorine can be added to water and the occasions when superchlorination may be required. After superchlorination, dechlorination is necessary to reduce the residual chlorine to about 0.5 mg l^{-1}, to prevent people complaining of the taste of chlorine in the water.

7.2.2   Kinetics of disinfection

The ability of a reagent to kill pathogens is related to the concentration of the disinfectant and the contact time. For chlorine, these factors are brought together in the equation

$$t^2 = \frac{2}{k} \ln \frac{N_o}{N_t}$$

However, this is not the general equation for all other disinfectants, which is:

$$\ln\left(\frac{N_t}{N_o}\right) = -kt$$

where

N_t is the number of viable organisms existing after contact time t

N_o is the number of organisms existing initially

k is the reaction rate constant for the disinfectant (s^{-1})

t is the contact time (s).

EXAMPLE 6

Ozone is to be used to obtain a 99.9% kill of bacteria in water with a residual of 0.5 mg l^{-1}. Under these conditions the reaction rate constant k is 2.5×10^{-2} s^{-1}.

Determine the contact time required.

ANSWER

$$\ln N_t/N_o = -kt$$

Therefore

$$t = \frac{1}{-k}\ln\frac{N_t}{N_o}$$

$$= \frac{1}{-2.5\times10^{-2}}\ln\left(\frac{0.1}{100}\right)$$

$$= \frac{1}{-2.5\times10^{-2}}(-6.908)$$

$$= 276.3 \text{ s}$$

EXAMPLE 7

What contact time will be required to give an *E. coli* kill of 99.99% for a free chlorine residual of 0.2 mg l^{-1}? Take $k = 1 \times 10^{-2}$ s^{-2}.

ANSWER

For chlorine,

$$t^2 = \frac{2}{k}\ln\frac{N_0}{N_t}$$

Thus for kill,

$$t^2 = \frac{2}{10^{-2}}\ln\frac{100}{0.01}$$

$$= 2\times10^2\times\ln 10000$$

$$= 200 \times 9.210$$

$$= 1842 \text{ s}$$

$$t = 42.92 \text{ s (say, 43 s)}$$

7.3   Other disinfectants

The Set Book discusses the use of chlorine dioxide, ozone and ultraviolet radiation. More information on the last two of these is given below.

7.3.1   Ozone

Ozone (O_3) is a triatomic form of oxygen, produced by passing dry oxygen or air through an electrical discharge. It is an unstable, highly toxic blue gas with a pungent odour. Its corrosive nature means that only materials such as stainless steel and polyvinyl chloride (PVC) can be used to handle it. It breaks down to give oxygen molecules (O_2) and highly reactive hydroxyl free radicals. The low solubility and instability of ozone means that it must be generated on site and introduced into the water as fine bubbles. A retention time of 4 minutes with a level of 0.4 mg l^{-1} is usually sufficient to destroy pathogens of concern in water.

For clear water, ozone is an effective way of destroying pathogens, and removing taste, odour and colour. It does not produce undesirable products, e.g. chlorophenols and THMs. The breakdown of ozone, however, means that it has no lasting residual; if this is required in the distribution system, it will be necessary to add chlorine.

7.3.2   Ultraviolet radiation

Ultraviolet radiation between the wavelengths 200–280 nm, peaking at around 262 nm, has been shown to have destructive properties against micro-organisms. It kills or inactivates viruses and bacteria by affecting their RNA and DNA, thus preventing their replication. The key operating parameters are the power intensity and the dose applied. There should be minimal suspended matter in the water being treated, as this can shield micro-organisms and prevent them from being irradiated. Scaling of the UV tube can occur with hard waters.

Water also absorbs radiation so the supply must be radiated over a shallow depth, and if suspended solids, colour, etc., are present, the UV dose must be increased proportionately. UV radiation has no residual effect and also forms no undesirable products. It is used for some small water supplies and in hospitals where high-quality water is required for specialist use.

7.4   Other forms of disinfectants

Substances such as bleach, potassium permanganate, silver bromide and iodine, have all been used in the disinfection of water. Boiling is a time-honoured way of destroying pathogens in water. Physically, micro-organisms can be removed using ultrafiltration and reverse osmosis.

7.5   On-site electrolytic chlorination (OSEC) systems

Fears over accidental leakage of chlorine gas has prompted a move towards the increased use of sodium hypochlorite. This can be obtained in bulk, or it can be generated through the electrolysis of brine. In the latter, a solution with a chlorine content of 6–9% is produced. About 3.5 kg of salt and 5 kWh of energy are required to produce 1 kg of chlorine. Potential problems with OSEC include the production of chlorates and bromates as undesirable byproducts.

SAQ 36

Which of the following statements on disinfection are true?

A During chlorination water should be maintained at a pH of below 7.0.

B The disinfecting agent formed by chlorine gas in water is the hypochlorite ion OCl⁻.

C Free available chlorine is the volume of chlorine gas dissolved in water.

D 'Combined available chlorine' refers to the monochloramines and dichloramines present in the water.

E If breakpoint chlorination is carried out, this ensures that the concentration of chloramines in the water will be high.

F Although expensive, chlorine dioxide may be used as a disinfecting agent since it does not combine readily with ammonia, it is stable, and does not form THMs.

G Superchlorination will remove all the sources which could give rise to taste and colour complaints, and at the end of the process produces a water ready for the public supply.

SAQ 37

What contact time will be required to give an *E. coli* kill of 99.9% for a free chlorine residual of 0.2 mg l⁻¹ (k value 10^{-2} s⁻²)?

SAQ 38

It is proposed to use a new disinfectant in water treatment. The disinfectant has a reaction rate constant of 3×10^{-2} s⁻¹. What will be the contact time necessary for this disinfectant to give a 99.9% kill of micro-organisms if the disinfectant follows the kinetics of a normal disinfectant?

8   INDUSTRIAL WASTEWATER TREATMENT

8.1   Introduction

Water sources can become polluted by effluents. In T210/T237 *Environmental Control and Public Health*, we considered sewage and its treatment, and here we concentrate on industrial effluents.

Industrial wastewaters or trade effluents tend to be characterised by great variability in flowrate and composition. The majority of industries are small to medium sized and do not operate continuously. Hence they do not produce an effluent continuously, or even one of consistent quality. It is only very large installations, such as refineries, that tend to produce a consistent quality effluent, due to their objective of producing a consistent quality product.

In small industries, a dairy, for instance, the effluent-producing operations such as tank cleaning are discontinuous, giving rise to variability of both flow and load for treatment. If the dairy manufactures different products (a range of milks, yoghurt, cheeses, etc.) the different processes will produce effluents of different volume and composition.

The short sewer lengths through which industrial wastewater is conveyed to an in-house treatment plant accentuates the variability in effluent quality. In a municipal sewerage system, the effluent would have been subjected to considerable mixing, dilution and attenuation, due to the extensiveness of the sewerage network.

Industrial effluents are likely to contain chemicals that are toxic, corrosive, or of extreme pH. Small amounts of these can be tolerated in a treatment plant but large quantities may disrupt treatment systems, especially if they are biological in nature. Consequently, an industrial wastewater treatment plant must be able to accommodate a range of flow and composition variations. There are also likely to be shock loads in the form of erratic discharges of specific chemicals or waste which may be dumped into the factory drainage system. Facilities for flow balancing and neutralisation (to achieve a uniform flowrate) are often included in industrial wastewater treatment plants.

SAQ 39

List some of the characteristics of industrial wastewaters which make its treatment more difficult than the treatment of domestic sewage.

8.2   Trade effluent control

There are several legal measures available to control pollution of the aquatic environment by trade effluents. For direct industrial discharges, strict consent conditions have to be met. These conditions may be such that expensive treatment plants may have to be installed before discharge can take place to a river or estuary. If the cost is high then it is often cheaper to discharge to the sewer and have the wastewater treated at the local sewage works, although this will also incur costs. This, of course, imposes a considerable treatment problem on the sewage works, which also has standards to meet before its own effluent can be discharged.

8.2.1   Discharge to sewer

The general principle that the best way for industrialists to dispose of their effluent is to discharge it to the sewer has been recognised in the UK since the early 1900s. The objective has been to ensure that the trader pays a charge for the services rendered for the reception, conveyance and treatment of the industrial effluent (the treatment required would depend on the consent conditions imposed on the sewage works). The principle was incorporated into the 1937 Public Health (Drainage of Trade Premises) Act, which allowed local authorities to recover some of the costs for the reception and treatment of industrial effluents. It has been pursued more forcefully by the 1961 Public Health Act, the 1973 Water Act which passed responsibility to the regional water authorities, and the 1974 Control of Pollution Act (COPA). The COPA gave control over discharges previously exempted by the 1937 Act and also allowed the conditions previously laid down to be varied. Now the water and sewerage companies issue consents to discharge in England and Wales. In Scotland, the 1968 Sewerage (Scotland) Act gave traders the right to discharge to public sewers, subject to compliance with conditions and charges imposed by the then Regional Councils. The latter duties are now the responsibility of Scottish Water. An example of consent conditions for trade effluent discharge was given in T210/T237 *Environmental Control and Public Health*. In Northern Ireland, Water Service issues the consents to discharge.

8.2.2   Discharge to rivers

The Environment Agency determines discharge consents to rivers in England and Wales. In Scotland, the Scottish Environment Protection Agency carries out the same function. Consents in Northern Ireland are determined by the Environment and Heritage Service.

8.2.3   Control of discharge to sewers

In 1976, the Confederation of British Industry (CBI) and the then Regional Water Authorities jointly produced a document entitled 'Trade Effluent Discharged to the Sewer: Recommended Guidelines for Control and Charging'. It was revised in 1986 and is shown in Appendix 5.

EXERCISE

Read Appendix 5 now.

This document covers the main factors on trade effluent charging. Since this is such an important document it is worthwhile highlighting the main points raised.

1   The objectives are fairly simple, namely to prevent damage to any part of the sewerage system, to protect involved personnel, to protect the treatment processes and to aid the water and sewerage company to foresee future demands. The most important economic factor is to ensure that the trader pays a fair charge. It is also stated that the water and sewerage company does not have to accept all industrial effluents and thus it may be necessary for the trader to carry out some pretreatment, to meet consent conditions, before discharge to the sewer. An example might be the presence of a high concentration of heavy metals which, if not all chemically removed by the trader, might not be completely removed by the subsequent treatment at the receiving sewage works. They could then enter the receiving stream, placing the sewage works in danger of not meeting its own consent conditions. The metals removed in the sewage treatment will appear in the sludge produced, giving rise to an environmental disposal problem.

2   Not all discharges are controlled and the exceptions are quoted. Note particularly the way 'Crown Privilege' exempts many establishments, such as ordnance factories, which could discharge undesirable compounds, and also the fact that radioactive substances are covered by the Radioactive Substances Act 1960. The following lists the properties enjoying Crown Privilege:

(a) land owned by the Sovereign;

(b) land owned by the Sovereign by right of the Duchy of Lancaster, Duchy of Cornwall, etc.;

(c) land owned by the Government through one of the Ministries.

3   As well as ensuring that the sewers are not affected by the industrial discharge, the consent should help the water and sewerage company to dispose of the contaminated sludge which can be produced in the sewage treatment process. Traders still have the right to reduce the strength of their effluent by treatment before discharge, and this could be directly to a river should this prove to be cheaper than discharging to a sewer. The trader will, of course, have to meet consent conditions set on this discharge.

4   Directions can be laid down to vary existing consents or to control discharges which were previously exempt.

5   Before charges can be calculated there are a number of factors to be considered:

(a) The treatment costs at sewage works. These can be calculated on a regional basis.

(b) Accurate measurement of the trade effluent volume. This can sometimes be difficult to achieve in old factories and, instead, the intake water meter reading can be used, with allowances being made for water used by the employees and any water used in the product, e.g. soft drinks.

6   The trade effluent is charged according to the formula

$$C = R + V + (O_t/O_s)B + (S_t/S_s)S$$

The origins of this formula (often referred to as the Mogden formula) can be traced back to 1936, when the first version was used to determine charges for the treatment of industrial wastewater at Mogden Sewage Treatment Works in the west of London. The terms are explained fully in the document but note that O_t is the chemical oxygen demand (COD) of the trade effluent in mg l^{-1} after the effluent has been allowed to settle for 1 hour with the pH having been adjusted to 7. This pH adjustment also applies to the total suspended solids in the trade effluent. The adjustment is carried out to achieve as near as possible the pH of normal crude sewage.

use COD as is quicker.

The biological oxidation cost is based on the COD and not, as one might have expected, on the BOD, as the COD test is simpler to carry out and can be completed in hours instead of days.

Looking at the formula, it can be seen that a reduction in trade effluent charges can be made by:

● reducing the volume of the effluent discharged

● reducing the contents of the effluent which contribute to the COD

● reducing the contents of the effluent which contribute to the suspended solids.

[Based on the above, a carpet factory in Kidderminster which uses wool (containing grease) found that it could dramatically reduce its trade effluent charges by using ultrafiltration to reduce the grease and suspended solids content in its effluent. A concentrated wastewater stream rich in potassium and nitrogen was generated and this was spread on grassland, while relatively clean filtrate was discharged to sewer.]

7   From paragraph 26 of the document it can be seen that the charge will only be levied for those elements of the formula which are applicable. Thus, if an industrial effluent is not biologically treated and the sewage works only carries out primary settlement, this part of the equation is ignored. Where the industrial effluent mixed with domestic sewage is discharged via a sea outfall, the charge levied can be the reception and conveyancing factor R, along with a special sea outfall charge, M.

8   A further point to note is that where a trader makes a capital contribution, e.g. to aid in the construction or extension of a sewage works so that the industrial effluent can be treated, then a reduction in the trade effluent charges should be made.

9   One final factor which must not be forgotten is that of monitoring. Monitoring is carried out for two reasons:

   (a) to ensure consent conditions are being met;

   (b) to provide an assessment of the strength of the industrial effluent and thus allow the charge to be calculated.

EXAMPLE 8

A trader discharges 1.4×10^3 m^3 d^{-1} of effluent containing 200 mg l^{-1} of chromium into the sewerage system. The dry weather flow (DWF) at the nearby sewage works is 60×10^3 m^3 d^{-1}. During normal operation, it is found that 70% of the chromium settles out in the primary tanks and is left in the sludge. The activated sludge process in the secondary treatment process removes a further 80% of the remaining chromium and 50% of this activated sludge is taken off as surplus sludge and mixed with the primary sludge for digestion.

(a) What weight of chromium per day will be present in the total sludge going to the digester?

(b) What will be the concentration of chromium in the final effluent?

(c) If the receiving stream gives a fourfold dilution and contains no chromium prior to the effluent discharge, what will be the chromium concentration in the stream?

(d) After receiving the effluent, could the stream be used as a source of drinking water supply?

ANSWER

(a) Discharge $= 1.4 \times 10^3$ m^3 d^{-1}

   Cr $= 200$ mg l^{-1}

   Weight of chromium discharged per day:

   $= 1.4 \times 10^3 \times 200/10^3 = 280$ kg

   But 70% of chromium appears in primary sludge:

   $$= 280 \times \left(\frac{70}{100} \right) = 196 \text{ kg}$$

   Therefore, chromium passing to aeration unit

   $= (280-196)$ kg $= 84$ kg

   In aeration unit, 80% of chromium removed

   $$= 84 \times \frac{80}{100} = 67.2 \text{ kg}$$

Half of this is taken off as surplus sludge, i.e. 33.6 kg.

Therefore, total chromium in the sludge sent to digester

$= 196 + 33.6 = 229.6$ kg

(b) Chromium in final effluent:

$= (84 - 67.2)$ kg $= 16.8$ kg

Works discharge $= 60 \times 10^3$ m^3 d^{-1}

Therefore, chromium concentration in final effluent

$= \dfrac{16.8 \times 10^3}{60 \times 10^3}$ g m^{-3}

$= 0.28$ g m^{-3}

(c) With fourfold dilution, chromium concentration in stream

$= 0.28/4 = 0.07$ g m^{-3}

(d) The stream would not be an acceptable source for drinking water supply. From Table A1 in Appendix 1, chromium must not have a concentration greater than 0.05 mg l^{-1}. Thus the trade effluent would have to have a lower value for chromium, if the stream were to be used as a water source.

EXAMPLE 9

A trade effluent from a sugar beet factory is discharged to sewer. The effluent has a COD of 1150 mg l^{-1}, a suspended solids content of 800 mg l^{-1}, and is discharged at a flowrate of 200 m^3 d^{-1}.

Answer the questions below, using the formula (included in the Water models on the Resources DVD)

$$\text{cost per m}^3 = R + V + \left(\frac{O_t}{O_s}\right)B + \left(\frac{S_t}{S_s}\right)S$$

where

$R = 13.17$p m^{-3}

$V = 12.74$p m^{-3}

$O_s = 542.8$ mg l^{-1} COD

$B = 18.55$p m^{-3}

$S_s = 347.6$ mg l^{-1} suspended solids

$S = 11.97$p m^{-3}.

(a) Calculate the total daily charge for discharging the trade effluent to sewer.

(b) A filter unit is planned as a pretreatment step. It removes all the suspended solids but leaves a residual COD in the effluent, depending on the rate of throughput used. Calculate the daily charge for discharging the pretreated effluent, for residual COD levels varying from 600 to 0 mg l^{-1}, and plot the results in a graph.

ANSWER

(a)

$$\text{Cost per m}^3 = 13.17 + 12.74 + \left(\frac{1150}{542.8}\right)18.55 + \left(\frac{800}{347.6}\right)11.97$$

$$= 92.76\text{p} = £0.9276$$

Total daily charge $= 200 \times 0.9276 = £185.52$

(b)

$$\text{Cost per m}^3 = R + V + \left(\frac{O_t}{O_s}\right)B$$

$$\text{total daily charge} = \frac{200}{100}\left(13.17 + 12.74 + \left(\frac{O_t}{542.8}\right)18.55\right)$$

$$= £\left[\frac{200}{100}(25.91 + 0.0342O_t)\right]$$

$$= £(51.82 + 0.0684O_t)$$

COMPUTER ACTIVITY I

The above expression (in the answer to Example 9b) can be evaluated for O_t values from 600 to 0 mg l^{-1} and a graph plotted, as shown in Figure 8. The activity is described in detail in the T308 Computer Activities supplement, and the spreadsheets you need can be found in the modelling section of the Resources DVD.

O_t /(mg l^{-1})	Daily cost (£)
600	92.86
500	86.02
400	79.18
300	72.34
200	65.50
100	58.66
0	51.82

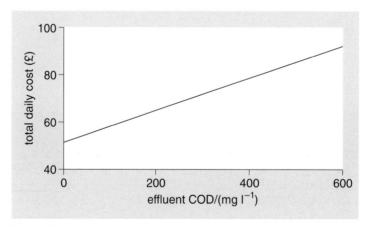

Figure 8

8.3   Trade effluent sampling

The important factors in trade effluent sampling relate to:

(a) estimation of flow

(b) estimation of the average value for the pH, COD and suspended solids.

The variability and the strength of the COD are also important.

Accurate flow measurement is easily obtained by using reliable flowmeters, but the problem of the variability of effluent strength is more difficult to overcome. It is expensive to obtain manually taken 'spot' samples, and automatic samplers can be extremely expensive. The question thus reduces to how many 'spot' samples are required to give an accurate reflection of the effluent quality at a reasonable cost.

For any set of analytical results, the number of determinations (n) required to give a result to a given precision is governed by

$$n = \left(\frac{z\sigma}{pm} \right)^2$$

where

n is the number of observations

z is the test statistic, dependent on the significance level of the test, e.g. for a 5% significance level it is 1.96

σ is the standard deviation of the data (the most common measure of the spread in a set of results)

p is the fractional percentage error allowed from the true result, e.g. for 10% error, $p = 0.10$

m is the arithmetic mean of the data.

EXAMPLE 10

How many samples will be required to obtain a result within $\pm$ 10% if the standard deviation is equal to one quarter of the mean and a 95% confidence limit is required?

ANSWER

$$n = \left(\frac{1.96 \times 0.25}{0.1 \times 1} \right)^2$$

$$= \left(\frac{0.49}{0.1} \right)^2$$

$$= 24.01 \text{ samples (say, 24)}$$

SAQ 40

What discharge consent for chromium would have to be imposed on the trader in Example 8 for the receiving stream to be suitable for abstraction of water for potable supply?

SAQ 41

The biological oxidation cost used in trade effluent charging is based on the COD and not the BOD. Suggest advantages and disadvantages in using these parameters.

SAQ 42

A trade effluent from a chemical plant is discharged to sewer at a flowrate of 300 m^3 d^{-1}. The COD and suspended solids content of the effluent are 1500 mg l^{-1} and 250 mg l^{-1}, respectively. Trade effluent charges are based on the following formula

Cost per m$^3 = R + V + (O_t/O_s)B + (S_t/S_s)S$

where

$R = 13.17$p m^{-3}

$V = 12.74$p m^{-3}

$O_s = 542.8$ mg l^{-1} COD

$B = 18.55$p m^{-3}

$S_s = 347.6$ mg l^{-1} suspended solids

$S = 11.97$p m^{-3}

sea outfall charge, $M = 9.59$p m^{-3}.

(a) Calculate:

(i) the total daily charge for discharging the trade effluent to the sewer

(ii) the total daily charge for discharging the trade effluent to sewer if uncontaminated cooling water equivalent to 30% of the flow is removed prior to discharge

(iii) the total daily charge if, in addition to the removal of the cooling water, the COD is reduced to 700 mg l^{-1} and the suspended solids to 50 mg l^{-1} prior to discharge

(iv) the charge if the trade effluent (in its entirety, i.e. 300 m^3) is discharged by sea outfall, rather than to a sewer

(b) There is a proposal to install a filtration unit to remove 95% of the suspended solids present in the original effluent.

(i) Plot a graph of the cost of discharging the pretreated effluent to sewer with varying levels of COD.

(ii) Determine the level to which the COD should be reduced to arrive at a cost of discharge which is half that of discharging the untreated effluent.

8.4   Waste minimisation

Waste minimisation is the continual process of reducing all forms of waste, preferably at source. It is a process that always pays dividends, very often at little or no cost. In T210/T237 *Environmental Control and Public Health*, we learned of several ways in which water savings could be made.

SAQ 43

Suggest reasons why waste minimisation is important with regard to water.

SAQ 44

Outline possible water-saving measures in the home.

One of the major uses of water in the modern home is in toilet flushing. Attention to this aspect has produced some alternatives to the water-based WC.

8.4.1   Vacuum system of sewage conveyance

Since conventional sewerage systems use enormous quantities of water, other methods of sewage conveyance down sewers have been tried. One such method is the Liljendahl system which uses air and a small quantity of water to transport the wastes from WCs down sewers. The air pressure within the sewer is maintained at approximately one half of the atmospheric pressure by means of a vacuum pump. Each operation of the WC admits about 18 litres of air together with the WC's contents and 1 litre of flushing water, and the discharge is conveyed as a plug by the excess upstream pressure resulting from successive flushings. The system involves power costs for the pump operation and an efficient maintenance service. But the conduits, usually of rigid PVC, are smaller than those of conventional systems and the quantities of wastewater are much less. One advantage of this system is that, since flow does not rely on gravity, the conduits can be laid on negative gradients where necessary, thereby avoiding costly excavation. This system is suitable for locations where a conventional system would involve high construction costs and where sullage (the wastewater from personal washing, laundries, food preparation) can be disposed of locally.

Sewage treatment on site

Sewage can be treated by composting at source, saving the expense of water and conveyance to treatment plants. Then only sullage needs to be despatched to the treatment works.

One possible composting system is the Clivus™ devised more than 60 years ago in Sweden.

8.4.2   The Clivus™ system

The pan in the Clivus system resembles the conventional type but is made from stainless steel and sits atop a stainless steel chute 35 cm across (Figure 9). Waste passes down the chute and lands on a bed of wood shavings contained in a large polyethylene tank, where it slowly degrades into compost. Liquid waste drains away and has to be collected for use or disposal.

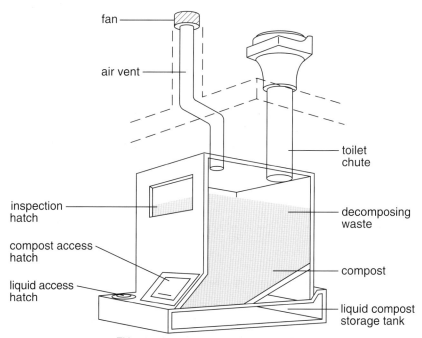

Figure 9   The Clivus™ composing toilet system [Courtesy of Kingsley Clivus Environmental Products Ltd]

Smell nuisance is minimised by having an electric fan which draws air through the compost pile. Aerating the pile discourages the growth of anaerobic bacteria which generate foul-smelling products such as mercaptans and other sulphur-containing compounds.

The Clivus toilet is said to require very little maintenance. The wood shavings act as bulking agents and help to create cavities, allowing air through the pile. The base of the composting tank slopes downwards and the compost slides by gravity to a collecting tray. Two litres of wood shavings are required to be put down the toilet each week.

The main disadvantage with the above system is that the toilet requires space beneath to a depth of at least 1.7 m to accommodate the tank, thus making it unsuitable for flats and high-rise buildings. To overcome this, a low-flush version of the toilet has been developed. This uses 0.5 litre of water to flush the contents of the pan to a tank set to one side. Liquid compost is removed once every 5–6 months, and solid compost is removed annually.

The major advantages of the Clivus system (introduced to the UK in 1991) are the savings in water usage and the absence of a pollutant discharge to waterways.

A variant of the Clivus composting toilet is used to dispose of both sewage and dry organic kitchen and garden wastes. These wastes are tipped in through a separate opening (Figure 10). Inverted U-shaped ducts and the ventilation pipe encourage the passage of air through the mass, preventing it from becoming anaerobic, and allowing excess moisture to evaporate.

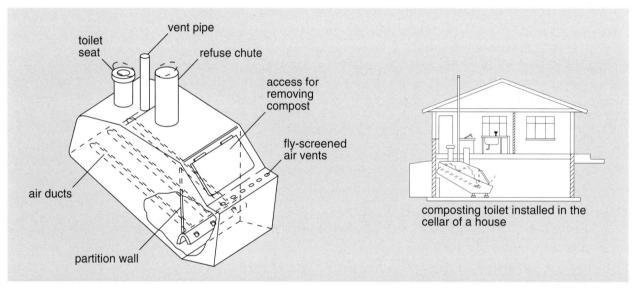

Figure 10   Continuous composting toilet for sewage and other organic wastes [Redrawn from Franceys, Pickford and Reed, 1992]

8.4.3   The waterless electronic biological composting toilet

This is another dry toilet system (Figure 11), also from Sweden. The sewage is heated to eliminate most of the water vapour, which is extracted together with the smell via a vent pipe fitted with an electric extractor fan. The remaining solid waste is rotated and composted. The temperature in the unit is maintained high enough to achieve pasteurisation of the solids.

The electricity consumption is said to be 120 W when the unit is used by a family of four. If heat from an external source (e.g. an internal combustion engine) is available, the electrical demand will be reduced to under 25 W. In outdoor areas, solar panels can be exploited to provide heat.

The final product from the toilet is said to be similar to peat moss. The toilet automatically empties this into a plastic bag (every 4 weeks for a family of

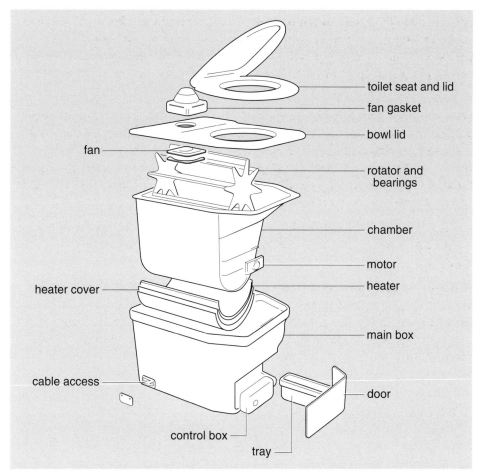

Figure 11   The waterless electronic biological composting toilet [AC Southwest]

four). No chemicals are used in the composting process and the final product is
said to be suitable for gardening use.

SAQ 45

Select the true statements from the following.

A   In the Liljendahl system of sewage conveyance, there is a total vacuum and
no water is required.

B   The best thing about the Liljendahl system is that it does not rely on
gravity.

C   The Clivus system is essentially a composting toilet but requires wood
shavings as a source of nutrient.

D   With the Clivus system, effluent flow is still present.

E   In the waterless electronic biological composting toilet, electricity is needed
to work the extractor fan only.

F   The Clivus and the waterless electronic biological composting toilet both
originate from Sweden.

8.5   Reducing wastage of water in industry

Industry has several options with regard to minimising waste of water. These
can be considered under the headings of: process modification; water reuse;
partial treatment and reuse; full treatment and recycling.

8.5.1   Process modification

A rethink of a given process can often result in a remarkable reduction in water usage. For instance, replacing wet cooling towers with air coolers can bring about a massive reduction in water usage and waste generation. Other simpler measures, often considered 'good housekeeping', can also be beneficial. An example would be the fitting of triggers to hoses in wash-down areas to help prevent wastage by errant workers. (As soon as the trigger is released, i.e. when the hose is put down, the flow ceases.)

Another example is the use of vacuum cleaners to remove spills, instead of using water to wash them away to a drain.

Water usage can be optimised with regard to the various operations in industrial practice. Table 15 summarises the possible options for batch and continuous processes.

Table 15   Optimisation of water usage in batch and continuous chemical processes

Process type	Batch	Continuous
Process control	Optimise mass balance	Control water to specific requirement
	Control raw material input accurately	Segregate waste streams
	Control reactor conditions, for optimum conversion	Minimise carry-over of chemicals after immersion of material in treatment bath
	Optimise phase separation	Replenish to prolong run times
	Slow down procedures	
	Reprocess off-specification material	
	Enhance reaction rates, e.g. by using catalysts	
Plant cleaning	Optimise campaign working, e.g. use reactor to produce a given product for as long as possible before switching to another product	Reduce frequency
		Use pressure techniques
	Recover product from first flush	Include automatic shut-off
	Reuse final rinse	
	Use clean-in-place techniques (these are water efficient)	
Support utilities	Recover spillage and detect leakages	Recycle or reuse cooling water
	Optimise cooling water management and steam usage	Optimise steam usage

It should be remembered that the workforce would also use water for hygiene purposes. About 50 litres per person per day is the norm. Anything much in excess of this implies that water saving measures, as applied in the home, are required.

8.5.2   Water reuse

Contaminated water from one process can be used in a second process, providing it is capable of accepting it. An example would be the use of boiler blowdown as washwater for filters. Figure 12 shows the hierarchy of water quality for different industrial uses. It is standard practice now in many plants to reuse the final rinse water as a pre-rinse.

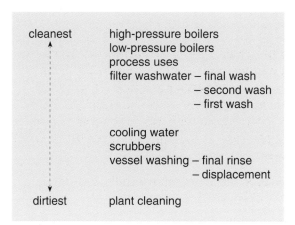

Figure 12   Hierarchy of water quality for different uses in industry

8.5.3   Partial treatment and reuse

Often partial treatment of an effluent can produce water of quality suitable for reuse. An example would be the reuse of plant cleaning water for irrigation, after removal of suspended solids and oil.

8.5.4   Full treatment and recycling

In certain circumstances it may well be worth treating an effluent fully and reusing it in the same process from which it was generated. This is especially so with processes requiring large quantities of water in arid regions of the world. As an example, in Muscat (Oman), there is a textile manufacturing plant which treats and reuses most of its process wastewaters. 'Most' because in order to prevent the build-up of trace components, a small proportion of the effluent has to be discarded and the process flow topped up with freshwater. Another example is in the microelectronics industry where spent rinse water (which is often of better quality than the mains water supplied to the factory) is treated with activated carbon to remove organics, and deionised before being reused.

8.6   Treatment options to enable reuse

Industrial effluents can contain a variety of contaminants which prevent their reuse. The most common pollutants and possible methods for their removal are presented in Table 16.

In all the processes in Table 16, clean water suitable for reuse may be produced, but the burden of disposing of the contaminant separated from the effluent will still need to be addressed, e.g. suspended solids separated from the effluent form a sludge which, if innocuous, can be landfilled. If the sludge contains toxic materials, however, a very different approach has to be applied, e.g. incineration, vitrification, etc.

In certain instances the contaminant may be transferred to another medium; for example, the removal of ammonia by air stripping relocates the ammonia from the liquid phase into the atmosphere.

Table 16   Possible treatment options for the removal of various industrial pollutants

Pollutant	Treatment option
Suspended solids, oils	gravity separation; coagulation, flocculation and sedimentation; flotation; membranes; filtration; centrifugation
Dissolved solids	reverse osmosis; electrodialysis
Dissolved organics	aerobic treatment; anaerobic treatment; chemical oxidation; activated carbon; membranes
Heavy metals	chemical precipitation; ion exchange using resins; reverse osmosis; electrodialysis
Ammonia	steam stripping; nitrification; ion exchange (using clinoptilolite); air stripping; breakpoint chlorination
Phenols	solvent extraction; biological oxidation; activated carbon; chemical oxidation

Some of the purification processes presented above may even add substances to the water, e.g. chemicals used as coagulants to remove suspended solids. This may render reuse difficult if it becomes contaminated.

A good source of information on waste minimisation in the UK is the Envirowise programme formerly the government's Environmental Technology Best Practice Programme (see http://www.envirowise.gov.uk). Two case studies from this programme are featured later in this section.

SAQ 46

The owner of a car-wash system in an arid country recently considered the possibility of treating and reusing the effluent from the system for washing cars.

(a) What options would you choose for the treatment system?

(b) What residues would there be to dispose of?

8.7 A waste minimisation programme

8.7.1 Waste audit

A waste audit is the first step towards a waste minimisation study at any site. It enables a comprehensive view of all the material flows and usages on the site to be obtained. This allows attention to be focused on specific areas where waste reduction, and consequently cost savings, are feasible. The process of waste auditing must have the material and moral support of senior management to be successful. There should be a commitment from senior management for implementation of the selected waste minimisation options.

The audit begins with the establishment of an audit team. For a site with many activities, personnel representing the different operations should form the basis of the team. This will help to increase employee awareness of waste reduction, as well as encourage input and support for the waste minimisation programme.

The audit proper begins with information-gathering on the various processes at the site. Process flow diagrams are then produced. These are best drawn using a computer-aided design (CAD) system, since a number of modifications may have to be made before the drawings can truly be regarded as a correct and full picture of the process in question. Any process which is complex should be broken down into component activities so that it can be understood.

Once a process flow diagram has been constructed, it should be annotated to show all the process inputs and outputs, together with an indication of the environmental media affected by any wastes generated.

A mass balance should be undertaken to ensure that all inputs are accounted for in the output streams. It is rare for the mass of outputs to be exactly equal to the mass of inputs (due to evaporation losses, leakages, spillages, etc.) but any large discrepancy should be investigated – it may be due to errors in measurement or, worse, to an undiscovered leak in the system!

The next step is to quantify the waste streams. This is a critical part of the study. There are several methods of obtaining quantitative data – the most straightforward is by direct measurement, e.g. using water meters or other techniques such as weirs, for measuring effluent flowrates. In other instances, mass balances may need to be carried out. Chemical analysis of the effluent streams also needs to be undertaken to determine composition. Measurements of flow and composition must be made over a range of operating conditions, e.g. start-up, full production, shut-down, and washing out.

Figure 13 shows a process flow diagram for a rail maintenance depot, with the principal waste streams marked.

In terms of water usage and disposal, an accurate drawing of the effluent pipework is essential. This will aid plans for the segregation of effluents to enable recycling and reuse. All drawings should be checked so that they show the actual situation on site. Confusion over the origin of various wastewaters is a major factor complicating the characterisation of effluents as a necessary preliminary to waste recycling and reuse. Site surveys also serve to highlight any leakages and overflows which may be present. These should be eliminated immediately.

8.7.2 Methodology

Water is used in industry for a variety of purposes – for cooling, as a cleaning medium in gas scrubbers, for washing and rinsing, for steam cleaning, and even for conveying material. For each use the audit should consider the condition of

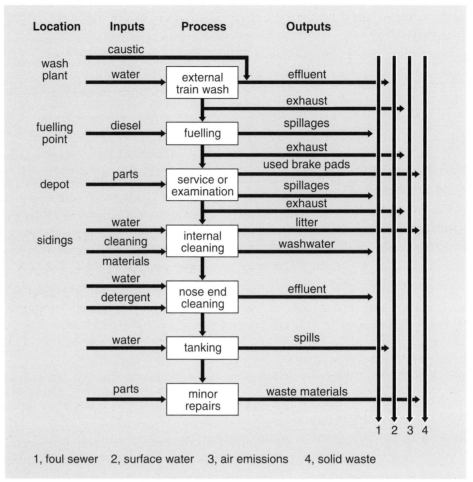

Location	Inputs	Process	Outputs

Figure 13 A process flow diagram for a rail maintenance depot [Courtesy of Nigel Clarke, Enviros Consulting Ltd]

water entering the process, its utilisation, and the nature of the effluent leaving the process. Examples of such considerations would include:

Incoming water

- What quantity and quality of feed water is required for each unit operation in the process?
- Are the quantity and quality more demanding than for similar processes in other plants elsewhere? (If so, a process modification may be warranted.)
- What sources of water are available on site? Are they of suitable quantity and quality for use, thereby eliminating the need to use the drinking water supply?
- What pretreatment may be required? Is the existing pretreatment system operating at its best? Can savings be made by using alternative methods or operating procedures?
- Has the water demand pattern changed recently or is it likely to do so in the future?

Production operations

- Are there ways of eliminating a water use (e.g. replacing cooling towers with air cooling units)?
- Can excessive demand peaks be avoided or reduced – for example, to improve consistency of water treatment, or to achieve water savings? Can this be achieved by minimising the rate at which water is used?
- Are there good-housekeeping measures that can be taken to reduce water usage?

- Is the water over-treated prior to use? Can lower-quality water be used satisfactorily?
- What is the rate of water usage for intermittent activities such as tank cleaning? (The rate can be very high!)
- Why is the effluent generated? Can entrainment of wastes in aqueous streams be reduced or eliminated?

Wastewater

- How many waste streams are there?
- Where do they go?
- Is it possible to segregate waste streams at source? Is it possible to recover materials from segregated streams using membranes?
- Can the effluent be reused in a unit process requiring low-quality water?
- How does the cost of pretreating the effluent compare with charges arising from discharge of the raw effluent to sewer?
- Would certain waste streams benefit from pretreatment prior to combination with other streams for end-of-pipe treatment?
- Would it be viable to treat the effluent for reuse?

Opportunities for saving water and reducing effluent flows will usually be revealed from this review. Combined with this review, if possible, a visit should be made to the plant when it is not operating. If done at night, when all should be quiet, the existence of water leaks could be revealed!

8.7.3   Option generation

Once waste streams have been targeted for reduction, waste minimisation options can be generated. These can range from minor process modifications (e.g. automating certain operations such as cleaning) to major changes (such as the conversion of the process from batch to continuous mode in order to reduce the amount of washing required). 'Cleaner' production technologies can be considered. These aim to minimise the generation of pollutants in a given process. A useful source of information on cleaner production is the International Cleaner Production Information Clearinghouse (ICPIC) established by the United Nations Environment Programme (UNEP) with the support of the US Environmental Protection Agency. ICPIC allows rapid world-wide access through fax and e-mail.

One effective way to generate waste minimisation options on site is to have a brainstorming session with personnel who have intimate knowledge of the process being considered. In brainstorming, the team tries to identify as many solutions to the problem as possible, however 'way out' they may seem. Table 17 suggests the rules to be followed in brainstorming.

Table 17   Rules for brainstorming

Generate as many ideas as possible
Do not be critical of ideas
Write down all ideas
Welcome creative thinking
Build on the ideas of others
Analyse the strengths and weaknesses of all ideas.

A second source of ideas and options for improvement is the workforce at each site. Maximising the involvement of staff increases not only the waste minimisation opportunities identified but also awareness and ownership of the project. Possible techniques for involving worker participation include suggestion schemes with financial rewards for effective options, and the use of existing quality circles in production units.

Input from workers is one of the best ways of identifying improvement opportunities for the numerous smaller (often occasional) sources of waste. Individually, these sources may be small but they can be so numerous as to make the overall cost of waste due to them very significant. They are unlikely to be identified by the audit team because their rate of waste generation is low, and are usually only known to the operators who work with the processes every day.

8.7.4   Evaluation

Each of the options generated above is evaluated for its technical and economic feasibility and its impact on the objective of minimising waste and conserving the environment. The outcome of these deliberations is a prioritised list of waste minimisation opportunities to be pursued.

Technical assessment
Pertinent questions include:

- Can this be done? Some solutions are just too far-fetched.
- Is this solution appropriate to the problem (for example, installing automatic controls where a simple change to operating procedures would be just as good)?
- Does it solve the problem or fix the symptom? For example, a solution that identifies a recycle route for waste merely deals with the waste – it does not reduce or eliminate it at source.
- Who can implement this? Is external help needed or can it be done in-house?
- When can it be implemented? Some solutions may take longer than others to come to fruition.

Any solutions that are deemed to be technically feasible now require economic assessment.

Economic assessment
An economic assessment of a potential solution to a waste minimisation problem includes:

- one-off cost of implementation (such as capital investment or work required)
- ongoing cost of operating or maintaining the solution (such as running costs, and inspection/auditing costs)
- the financial benefits from improved environmental performance, e.g. less charges paid out for effluent discharges to sewer.

In many cases a simple payback calculation is sufficient to assess the economic feasibility of a solution or to identify an optimum solution – the one with the lowest total costs for the business. Solutions that have a payback period of two years or less would be considered worthwhile implementing.

Calculating payback period

The payback period can be calculated by dividing the total one-off cost of the project by the net saving of the project each year (the difference between the gross saving and running costs). This gives a payback figure in years.

For example, a £42 000 piece of equipment with running costs of £8000 per year will save £67 000 in effluent charges annually. So the payback period is:

One-off cost (£42 000) divided by the net annual saving (£59 000) gives a payback of 0.71 years or 8.5 months.

Environmental assessment

It is very important to assess the effect that any planned changes have on the environment:

- Will the environment benefit? Consider the environmental effects of proposed changes. It is important to ensure that the overall environmental burdens are minimised.

- Is the environmental problem being transferred elsewhere? For example, replacing a wet scrubber with an air stripper does not fix the problem.

- Are there any environmental reasons for not proceeding with the project? Could legislation affect its desirability in the future?

Some companies are carrying out life-cycle assessments of their products and processes to identify those with the lowest environmental burden or impact.

Life-cycle assessments are discussed in the Wastes block.

8.7.5   Implementation of waste minimisation measures

Through evaluation of technical, economic and environmental factors, viable options are selected for implementation.

Some solutions will be straightforward changes to operating procedures or simple technical changes to equipment. Other projects may involve changing peoples' attitude and behaviour. Individuals will need to be more responsible for the consequences of their action or inaction. Environmental, health and safety, and business or costs awareness training could be a good way of delivering this information.

Capital investment projects will need planning and integrating into the business plan, taking into consideration when the capital and manpower to manage and implement the project will be available. Training may be needed.

Once implemented, the project must be monitored in order to confirm the return on investment and to learn from any mistakes. Above all, a successfully implemented project helps to maintain momentum for waste minimisation – 'Nothing promotes success like success'.

On commissioning the above changes, the plant database is updated.

8.7.6   Continuous monitoring and targeting

Setting up an effective monitoring programme is an integral part of an ongoing programme of waste minimisation. Once waste minimisation options have been implemented, it is necessary to measure the actual impact of the changes. Monitoring provides a measure of the effectiveness of the implemented options, and also helps to identify if any of the options have led to an adverse effect elsewhere in the system in terms of waste generation. This may not have been recognised at the time of planning or implementation.

By providing data on the outcome of implemented options, monitoring enables decisions to be taken on future priorities and on targets for further improvement, making waste minimisation an ongoing process of continual improvement. Indeed, waste minimisation has been described as 'a journey rather than a destination'!

8.7.7   Dissemination of results

The achievements and benefits of the waste minimisation programme should be publicised to all employees. This helps to motivate further efforts to reduce waste and improve performance, as well as greatly contributing to company image.

8.7.8   Review audits

It is sensible to set up a programme of review audits to ensure that waste minimisation is being continually applied. Such audits also provide the opportunity for regular re-evaluation of processing methods and procedures in terms of the latest technology, and can result in further changes as appropriate.

8.7.9   Manuals for waste audits

For a successful waste audit, an auditing manual and audit checklists are vital. A useful manual is the one published jointly by the United Nations Environment Programme (UNEP) and the United Nations Industrial Development Organisation (UNIDO), called *Audit and Reduction Manual for Industrial Emissions and Wastes*. The document, in addition to giving a step-by-step guide to auditing, illustrates the application of waste auditing and the waste minimisation measures arrived at in three different cases, involving beer production, leather production, and the manufacture of printed circuit boards.

Proprietary software is also available for the methodical preparation and execution of waste audits. These are particularly useful in complex plant situations where one can easily get confused with myriad process flows and interconnected systems.

SAQ 47

List the main steps in carrying out a waste minimisation programme specifically for water.

8.8   Case studies on water reuse and waste minimisation

The following are two case studies which illustrate water reuse, and the savings that were made possible.

8.8.1   A cheese-making plant in Cheshire

In cheese-making, milk is first pasteurised, and then a special bacterial starter culture is added to convert some of the lactose in the milk to lactic acid. Rennet is then added to the milk and within a short time a curd is produced.

This curd is cut into small cubes, and heat is applied to start a shrinking process, which, with the production of lactic acid from the starter cultures, changes it into small rice-sized grains. These are allowed to fall to the bottom of the cheese vat, and the remaining liquid, called whey, is drained off. The curd grains mat together to form large slabs of curd. The slabs are milled, and salt is added to provide flavour and help preserve the cheese. The slabs are later pressed, and subsequently packed in various-sized containers for maturing.

About 90% of the milk used for cheese-making ends up as whey, a liquid containing proteins and sugars (Table 18). This is valuable as animal feed. Whey is also used as an additive in foodstuffs such as ice cream, confectionery, and bakery products.

Table 18   Typical composition of whey

Component	Percentage
Water	93.7
Lactose	4.5
Protein	0.80
Minerals	0.75
Other solids	0.15
Fat	0.10

[Courtesy of Envirowise]

In many cheese-making plants this whey is tankered away to farms for use as animal feed, or to be spread on farmland. It is sometimes sent to effluent treatment works for disposal. In one particular cheese-making plant in Cheshire, in an effort to gain an income from the whey, as well as to secure a long-term disposal route for the substance, a whey recovery system based on membrane technology was installed. The recovery system consists of spiral-wound ultrafiltration, nanofiltration and reverse osmosis modules, in food-grade stainless steel housings. The plant is computer controlled to minimise energy consumption.

In the first stage, pasteurised whey is separated and concentrated by ultrafiltration to give a 50% whey protein concentrate. The ultrafiltration permeate is then passed through reverse osmosis and nanofiltration membranes to produce a lactose milk sugar concentrate with a 24% dry solids content. The whey protein concentrate and lactose milk sugar are stored in silos and sold as high-quality products. There are plans to install an evaporator and drier to concentrate and dry the recovered protein and lactose to further increase their saleable value.

The permeate from the reverse osmosis and nanofiltration units is then treated further via another reverse osmosis membrane with a smaller pore size to produce demineralised water, and a small amount of low-grade water. The demineralised water is used as boiler feed, or for cleaning the membranes, and the low-grade water for yard washing. Figure 14 shows the daily flows through the whey recovery system.

The membrane units are used 10 hours a day, 5.5 days a week, in line with cheese production. Whey processing is followed by automatic cleaning-in-place (CIP) of the membrane modules, which takes a further 3 to 4 hours. Spent cleaning-in-place solutions are discharged to the site's effluent treatment plant.

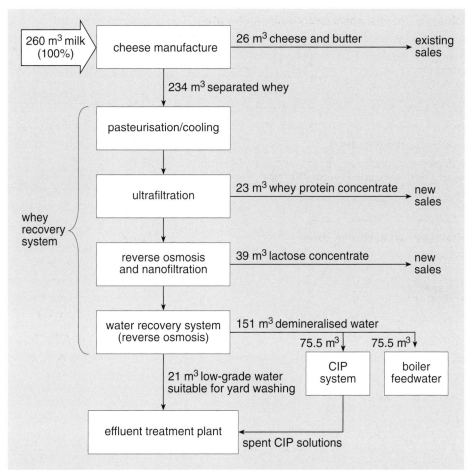

Figure 14   Flow diagram of the whey recovery system [Courtesy of Envirowise]

Cost benefits

Income of over £920 000 per year is now obtained from the separated whey products, compared with £18 600 previously. All the water needed for membrane cleaning is provided by the demineralised water from the water recovery system. The remaining demineralised water is used as boiler feedwater. Overall, the plant is self-sufficient in water, and provides a small surplus for other site uses.

Table 19 gives an economic analysis of the whey recovery system. The £1.02 million capital cost includes expenditure on buildings and services designed to accommodate long-term plans for an evaporator and drier. However, the net cost savings of over £685 000 per year still gave a payback period of only 1.5 years.

Other benefits

A major benefit of the whey recovery system is the removal of potential business constraints. Production peaks can now be accommodated readily, as whey treatment and disposal is controlled by the company. The risk of surface run-off pollution from spreading whey on farmland has been reduced, as have local tanker vehicle movements and their associated nuisance.

Table 19   Annual costs and savings at the plant

	Cost (£)	Saving (£)
Annual operating costs		
Electricity	100 000	
CIP chemicals	45 000	
Labour (2 full-time staff)	40 000	
Membrane replacement (based on full replacement every two years at a cost of £70 000)	35 000	
Wastewater treatment and disposal	4 200	
Avoided whey transport costs, etc.	(1 400)	
Total annual operating costs	**222 800**	
Annual cost savings		
Sale of lactose		168 000
Sale of whey protein		760 000
Loss of revenue from whey sales to farmers, etc.		(18 600)
Total annual cost savings		**909 400**
Net annual cost savings		**686 600**
Capital costs		
Membrane equipment	525 000	
Boilers, silos, pipework, etc.	495 000	
Total capital costs	**1 020 000**	

[Courtesy of Envirowise]

SAQ 48

(a) Calculate the volume of effluent treated annually, if the plant operates for 50 weeks a year.

(b) Calculate the payback period for the whey recovery plant.

8.8.2   A timber products plant in North Wales

The manufacture of medium density fibreboard (MDF) generates large amounts of effluent with a high chemical oxygen demand (COD) and suspended solids content. Previously, in a plant in North Wales, this was tankered off-site for treatment and disposal. This operation was expensive, and MDF production was interrupted if the tanker was late. The company considered various options, and decided on a hybrid system (comprising coagulation–flocculation, filtration, and reverse osmosis) for material recovery and reuse of water. The company negotiated a lease-purchase agreement under which the supplier of the system operates it for a monthly fee that is about half the cost of off-site effluent disposal.

In MDF production, forest thinnings and sawmill residues are debarked and chipped before being washed to remove sand, etc. The fibres are then softened using steam. Excess water is removed by a screw press before the chips are cooked at 70–80 °C. The chips are then 'refined' between two flat plates, and passed through heated cyclones to drive off most of the remaining moisture. The fibres are then poured on to a conveyor and premixed resin is added. The heated mix is passed between two moving steel belts to compress it to the desired thickness. The formed boards are then cut to the desired dimensions and stacked to cool.

Water is used in large quantities for chip washing and refining, steam raising, resin make-up, and in air pollution control. Most of the 48 000 m^3 of effluent generated in a year comes from the chip washing and refining stages. The effluent typically has a COD of 15 000 mg l^{-1}, and a suspended solids content of 5000 mg l^{-1}. The organic components in the effluent include celluloses, lignins and resin acids.

The high COD and suspended solids content of the effluent meant that it could not be discharged to sewer without treatment. It was tankered to an effluent treatment plant. Holding tanks for one day's storage exist in the MDF plant. In order to avoid problems due to delays in tankering, the company decided to search out a low-risk, low-cost solution that provided maximum wastewater reuse and material recovery.

Pilot trials of possible options were undertaken. The schematic of the system finally chosen is shown in Figure 15.

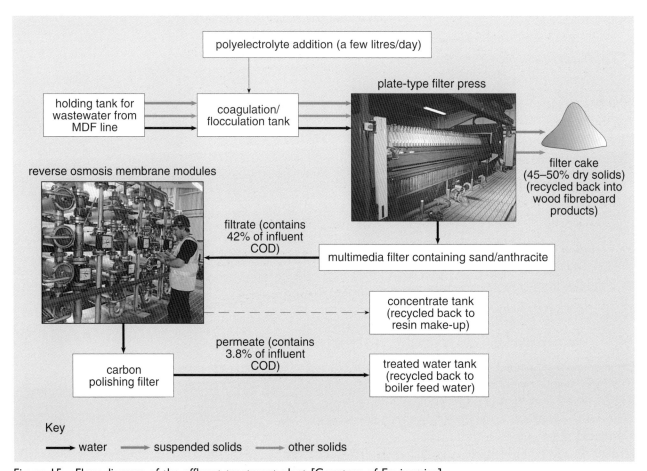

Figure 15   Flow diagram of the effluent treatment plant [Courtesy of Envirowise]

In the coagulation–flocculation and filtration stages almost 98% of solids greater than 5 μm in diameter are removed. This protects the reverse osmosis membranes and reduces the need for backwashing. Most of the dissolved organics are removed by the three-stage crossflow reverse osmosis membranes, thus reducing the COD to less than 10% of that of the raw effluent. Suspended solids are virtually all removed. To achieve the quality needed for reuse, the permeate from the membranes is passed through a carbon filter. The process is computer controlled, and average performance data are shown in Table 20.

Table 20   Average plant performance data

	Water (t d^{-1})	COD (mg l^{-1})	Suspended solids (mg l^{-1})
Feedwater	129.6	16 220	5 870
Filter cake	1.6	N/A	N/A
Premembrane filtrate	128.0	6 880	130
Concentrate	7.0	Not measured	130
Permeate	121.0	200–300	Negligible

[Courtesy of Envirowise]

The spiral-wound membranes used have a smooth finish to reduce surface build-up. They are chemically robust to allow high-temperature cleaning at low and high pH to remove surface build-up when it appears.

Approximately 93% of the feedwater ends up as clean permeate, and is used to provide some 60% of the boiler feedwater requirement. The concentrate emerging from the membranes consists of useful dissolved organics (including cellulose and lignins) and is added to the MDF resin binder. The filter press cake provides about 480 tonnes of dry solids a year, as a substitute feedstock for the wood fibreboard plant. The water mass balance for the plant is shown in Figure 16.

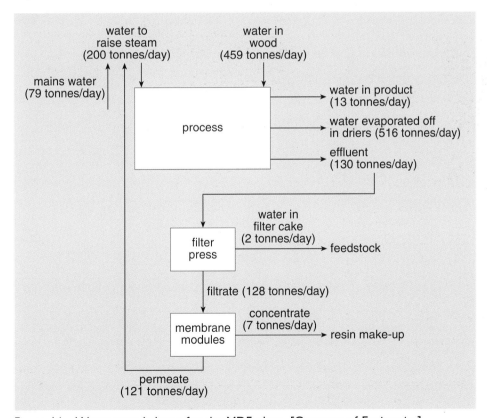

Figure 16   Water mass balance for the MDF plant   [Courtesy of Envirowise]

Cost and environmental benefits

The company pays £22 000 per month to the supplier to run the plant but even so savings are made (see SAQ 49 below). The new plant has eliminated the need to tanker effluent off-site for disposal, and reduced mains water consumption. All the effluent is reused, and using the concentrate as an additive in the resin binder has actually improved product quality.

SAQ 49

Use the figures in Table 21 below to calculate the net annual cost saving.

Table 21   Annual savings summary for the new treatment plant

	Cost savings ($£$ year^{-1})
Avoidance of need for tankering of 47 800 m^3 effluent off-site	468 500
Reduction in mains water consumption by 44 000 m^3	29 480
Replacement of raw materials with 480 tonnes recycled fibre	17 760
Sub-total of cost savings	
Less cost of fees to supplier-operator of new treatment plant	
Net annual cost saving	

9   ANALYSIS OF WATER

Except for simple on-site tests, it is likely that most samples at a water treatment plant will be taken to a centralised laboratory for analysis. At the main laboratory it is possible to have sophisticated equipment to determine trace concentrations of substances and to have a large degree of automation to reduce staff costs, and improve on the time taken to complete the required analyses. The laboratory can also incorporate the analytical control procedures which are necessary to give the required confidence in the results. It is likely that the laboratory would be able to analyse the water for a miscellany of chemical, bacteriological and biological parameters. Samples of water would be sent to the laboratory for a variety of purposes and the analyses performed would reflect the reason for the sample being taken. The data are also likely to be held on a complex computer database, with remote access from many sites, so that the information can be used for assessing trends, compliance with regulations, problem solving and for designing new treatment plants.

When a new treatment plant is planned it is necessary to know the main characteristics of the water to be treated. For example, the variations in the quality of a proposed water source must be recognised to determine the type of treatment likely to be required, which would affect the design of the plant.

At a water treatment works it is likely that some key parameters will be monitored continuously and the results fed into a telemetry system to give alarms when values differ from the norm. Chlorine, pH and turbidity are commonly determined at all sites together with other parameters such as fluoride, if this is artificially added. The simple on-site tests would be for optimising the plant operation and troubleshooting.

9.1   Sampling

To be of value, the results of any analysis must accurately reflect the condition of the water being sampled. The problems associated with obtaining a representative sample relate to such variables as flowrate, seasonal and diurnal variation in water quality, changes in the composition of polluting matter, etc.

Some of the problems may be overcome by taking individual samples on a time-related basis and combining these samples in proportion to flowrates into a composite sample. Composites of this type indicate average conditions but will not show individual pollution incidents. One way to overcome this problem is to use ion-selective electrodes permanently immersed in the flow which will give continuous monitoring of basic parameters, such as pH, dissolved oxygen, etc., and also to take individual samples for more complete analysis.

The problems associated with sampling are recognised by the EU directives which offer advice on the main features to be considered.

1   Sampling points. According to the Drinking Water Directive, samples should be taken where the water is made available to the user, while the Water Abstracted for Drinking Water Directive requires water to be sampled at the abstraction point.

2   Storage and preservation. The directives require that storage and preservation of samples should not significantly affect any analytical result. The Directive Concerning the Quality Required of Surface Water Intended for the Abstraction of Drinking Water states that glass containers should be used for phenols, hydrocarbons, pesticides and dissolved oxygen estimations, with sterilised glass bottles for microbiological samples. Nowadays, many sample containers have trace quantities of 'fixing agents' or preservatives to

minimise changes of parameters during transit to the laboratory. For example, dilute acid is used to ensure that metals remain in the water sample, and do not deposit on the inner surface of the glass, and caustic soda is used to stabilise cyanide complexes. When sampling for lead at domestic taps, the 'first runnings' must be collected to check for plumbosolvency. For assessment of organic chemicals, ultraclean borosilicate glass must be used.

The sterilised bottles for microbiological samples will contain a few drops of sodium thiosulphate solution to neutralise residual free chlorine immediately the sample is taken. Samples should be stored at a low temperature (4 °C) to reduce changes in composition during storage, and transportation to the laboratory.

All samplers employed by water companies in England and Wales are highly trained and certified. Likewise all analytical methods used are accredited and subject to analytical quality control.

3   Frequency of sampling. All the directives specify the frequency with which samples should be taken for some specified parameters.

The Directive Concerning the Quality Required of Surface Water Intended for the Abstraction of Drinking Water gives the monitoring frequencies for water requiring different degrees of treatment, i.e. A1, A2 or A3. The determinands quoted are divided into three groups and the frequency for each group depends on the population served. The most frequent monitoring required is monthly and this relates to water requiring substantial treatment for use by a large population.

The Drinking Water Directive uses the volume of water supplied to calculate the minimum frequency of sampling and analysis required (see Appendix A4). There are two types of monitoring: check monitoring and audit monitoring. In check monitoring, the purpose is to provide information on the organoleptic (relating to apprehending by the senses, namely taste, smell, colour, and appearance) and microbiological quality of the water, as well as information on the effectiveness of the treatment used.

The following parameters must be subject to check monitoring but Member States can add others if they deem it appropriate:

Ammonium

Colour

Conductivity

Clostridium perfringens, including spores (only if the water originates from or is influenced by surface water)

Escherichia coli

pH

Iron (only when used as a flocculant)

Nitrate (only when chloramines are used in disinfection)

Odour

Pseudomonas aeruginosa (only for water offered for sale in bottles or containers)

Taste

Colony count at 22 °C and 37 °C (only for water offered for sale in bottles or containers)

Coliform bacteria

Turbidity.

The purpose of audit monitoring is to provide information on whether or not all of the directive's parametric values are being complied with. Thus, in audit monitoring all the parameters are analysed for.

EXAMPLE 11

If 9000 m^3 d^{-1} of water is supplied in a small town determine the minimum number of samples required for check monitoring and audit monitoring.

ANSWER

From Table A5 in Appendix 4, for 9000 m^3 d^{-1}, for check monitoring a minimum of

$$4 + \frac{3(9000)}{1000} = 31$$

samples are needed each year.

For audit monitoring, a minimum of

$$1 + \left(\frac{9000}{3300} \right) = 1 + 2.73 = 1 + 3 \text{ (rounded up)} = 4$$

samples are needed each year.

SAQ 50

Determine the number of check and audit monitoring samples required in a year for a water supply of 110 000 m^3 d^{-1}.

SAQ 51

What are the main problems to be overcome in sampling for chemical and microbiological analyses?

9.2   Remote monitoring of water quality

With rapid advances in micro-electronics and telemetry, remote monitoring of water quality has become practicable. Remote water quality monitoring can provide early warning of pollution and can be used to monitor compliance with discharge consents. Monitors can be general or specific in nature. Examples of each type are described below.

9.2.1   General pollution monitors

Historically, water quality monitoring has been confined to the determination of specific parameters, e.g. dissolved oxygen, BOD, COD. In recent years there has been an increased interest in the use of toxicity as a measure of the pollution of a watercourse. The variety of potential contaminants being vast, a broadband pollution indicator is desirable. One such system is the WRc Fish Monitor (described below), designed for use at drinking water abstraction points in order to give early warning of the presence of potentially harmful chemicals in the water source.

The WRc Fish Monitor

This system is based on the fact that fish are very sensitive to changes in water quality, with the rate of breathing being quickly affected by contamination of the water. Developed by the Water Research Centre (WRc) in the UK in the early 1980s, the WRc Fish Monitor uses eight small rainbow trout in individual tanks (Figure 17), with non-invasive electrodes attached, in order to monitor the tiny electrical signals generated by the gill muscles of the fish during breathing. A constant flow of sample water from the site being monitored is passed through each tank at a rate of 2 litres a minute. Every minute, data are gathered from the fish and analysed by computer. By comparing current data against historical breathing rates, and using various statistical packages, the computer can make decisions with respect to water quality changes. The fish are kept for a week at a time in the tanks, and are not fed during this period as their metabolic reactions would interfere with the monitoring process. Increased fish activity can trigger the collection of a sample for subsequent analysis at the laboratory.

Figure 17   The WRc Fish Monitor

In practice, two sets of fish tanks (of four each) and the computer system are housed in three separate rooms. The fish are kept isolated to prevent them from being disturbed by external factors, such as human body movement, and to enable four fish to be changed while the other four continue their monitoring function.

In addition to the data generated by the fish, the computer can also accept signals from up to 16 other specific water quality sensors (discussed in the next section) and compare these with preset thresholds. The system can detect both gradual and rapid deterioration in water quality and raise alarms automatically by telemetry to contact a manned station.

Other systems for broad-spectrum pollution monitoring have also been developed. These typically involve monitoring fish behaviour and activity of bivalves (mussels). The Microtox™ and Amtox™ systems (introduced in T210/T237 *Environmental Control and Public Health*) also have the same function but use micro-organisms.

Fish monitors are used at some abstraction points on major rivers that receive inputs from industry and agriculture (mainly as diffuse pollution), and where well-used trunk roads have bridge crossings in fairly close proximity upstream of the intake. Not all water utilities use fish monitors at present. Many have been replaced by specific analytical methodologies that can be adapted for on-site monitoring stations. The activity ratios of captive fish are, however, still regarded as important indicators of the presence of unknown pollutants. Increases in fish activity trigger the taking of samples, and alarms that can be evaluated by the water abstractor.

The Watersure On-line Pollution Monitor

This is a general gross organic pollution monitor designed for use at water intakes. It is said to respond to lower levels of some pollutants than does the WRc Fish Monitor.

It is based on the principle that many organic molecules absorb light of a given frequency or range of frequencies. Thus any water sample which contains organic molecules will absorb more light energy (at given wavelengths) than a corresponding sample with no organic molecules present. The absorbance of light over a range of wavelengths (rather than at a single wavelength) is measured because:

(a) not all organic molecules absorb light at the same wavelength;

(b) not all organic molecules which absorb at a given wavelength have their maximum absorbance at that wavelength.

In order to minimise the effect of extraneous factors such as nitrate, hydroxide and carbonate ions and colour on the absorption observed so that any absorption is attributed only to organics present, a wavelength range of 250–290 nm (in the UV range) is used in the Watersure system.

To use the system, reference scans of the clean river water at various times of the day are first obtained and stored in a computer. This is necessary as the characteristics of the river will vary throughout the day. As well as diurnal variation, and other changes brought about by humans, prevailing weather conditions such as flash storms and prolonged dry periods can also have an effect. Then the monitor is activated. Samples of river water are collected and compared with the reference scan of the clean river sample. Using a complex mathematical and statistical package, the variation between the two samples is analysed. If found to be significant, the system diagnoses it as pollution and an alarm is raised.

9.2.2   Specific substance monitors

The Organic Pollution Alarm (OPAL)

This system (Figure 18) is used to detect the presence of volatile organic chemicals, such as solvents, and can be installed at water intakes. Clean air is bubbled through a water sample and any volatile organics present pass into the air, which is then directed into a photo-ionisation detector (PID). In the detector, UV light splits the organic molecules into ions. The ions allow an electric current to flow across the detector and this correlates with the amount of organics present.

The system responds well to most common solvents, especially toluene, petrol, etc., but the actual reading for each compound depends on its solubility in water and the response of the PID. The main function of the unit is to raise an alarm when a solvent is detected. The response time is less than a minute in most cases, and detection down to the μg l^{-1} level is possible.

Biosensors

In T210/T237 *Environmental Control and Public Health*, the Amtox system for on-line monitoring of nitrification inhibition (and hence toxicity) was described. Here, we consider biosensors in general.

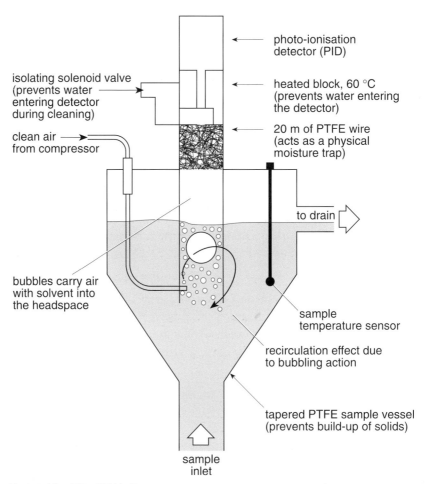

photo-ionisation
detector (PID)

isolating solenoid valve
(prevents water
entering detector
during cleaning)

heated block, 60 °C
(prevents water entering
the detector)

clean air
from compressor

20 m of PTFE wire
(acts as a physical
moisture trap)

to drain

bubbles carry air
with solvent into
the headspace

sample
temperature sensor

recirculation effect due
to bubbling action

tapered PTFE sample vessel
(prevents build-up of solids)

sample
inlet

Figure 18   The OPAL System

Biosensors involve the intimate connection of biological matter with a transducer which gives an indication of the rate of metabolic activity in the former. Biosensors were initially developed for medical purposes (e.g. for the determination of glucose) but are now used in a variety of fields such as in the food industry and the environmental, defence, industrial processing, agricultural, horticultural, and aquacultural sectors.

Whole cell organisms or organelles (parts of an organism) can be used as biosensors. Biosensors are portable and react rapidly to a wide range of toxicants. The toxic effect of a chemical can be deduced by the effect on the organism. For instance, the respiration rate of *E. coli*, or the rate of photosynthesis of *Synechococcus*, can be used to monitor the toxicity of given chemicals.

In a practical system, water requiring monitoring would be pumped through a flow cell containing, say, *Synechococcus* immobilised and preserved on working electrodes. The photosynthetic activity of the organism in clean water is first stimulated intermittently by light emitted from light-emitting diodes. The result of this intermittent stimulation is a series of peaks in electrical signals (Figure 19).

The sample water is then pumped through the flow cell. A decrease in the peak height indicates the presence of a chemical species affecting the biological process of photosynthesis. Figure 19 shows the drop in peak height when the organism is exposed to linuron, a pesticide. An on-line system can typically be operated continuously for up to one week before the electrode requires replacement.

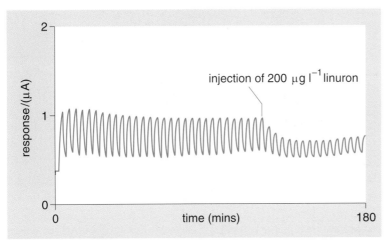

Figure 19   The effect of linuron on a biosensor based on *Synechococcus* [Courtesy of WRc, Swindon Plc]

A field kit based on the above principle has also been developed to assess the toxicity of discrete water samples on-site. The test organism is immobilised and preserved on a screen-printed electrode with a combined reference electrode and housed in a vial. The water sample is put into the vial and the change in current amplitude is used as a measure of toxicity. The unit gives a response within 20 minutes.

Trials undertaken by WRc reveal that the sensitivity of different organisms to different chemicals varies significantly – for example, *Synechococcus* reacts strongly with herbicides such as atrazine and bromoxynil, while *E. coli* is more sensitive to compounds such as tributyltin chloride and metals like mercury and copper. Thus to achieve effective broadband monitoring of pollution, a variety of species deployed in an array of biosensors should be used.

SAQ 52

Select the false statements from the following.

A With the WRc Fish Monitor, a change in the breathing rate of the fish is taken immediately as an indication that a deterioration in water quality has occurred.

B The fish in the WRc Fish Monitor are fed daily so as not to expose them to undue stress.

C In the Watersure On-line Pollution Monitor, the absorbance of light in the wavelength range 250–290 nm is measured in order to minimise the effect of any inorganics.

D With the Watersure system, water samples are compared with a 24-hour composite of clean river water.

E The Organic Pollution Alarm (OPAL) is a device which quickly detects the presence of volatile organics.

F Only whole cells can be used in biosensors.

G Because organisms differ in their sensitivity to different toxic compounds, a variety of biosensors, with different microbial species, would be ideal for wide-spectrum pollution monitoring.

9.3   UK emergency procedures to protect water quality

9.3.1   Introduction

While all water companies do their utmost to produce wholesome water for public supply, they do face potential problems from many different types of emergencies, which can affect the water supply both in terms of quantity and quality. For instance, there were major incidents of *Cryptosporidium* contamination in water supplies in England in 1989 and 1997, and in Scotland in 2002.

There was a serious case of a solvent contaminating the water supply in 1994 in Worcester. Going further back, a major incident occurred in 1988 with the accidental addition of aluminium sulphate to the water supply in Camelford.

While these incidents have been major, most incidents are relatively minor, affecting only a few consumers and therefore receiving very little publicity. The disruption could be something like a burst distribution pipe, which means that a small area does not receive any mains water for a few hours while repairs are carried out. In terms of quantity, the loss of water due to a burst pipe can sometimes be identified by the water company because of a pressure loss, or can easily be recognised by the customer and soon notified to the water company. Water quality problems are more difficult to identify by the customer unless the changes affect the physical senses of smell, taste or visual appearance. Even with sophisticated laboratory facilities, the water company cannot analyse water samples for all potential pollutants on a routine basis as the water goes from the treatment works to the tap, due to practical considerations and limited resources.

9.3.2   Water quality considerations

This section will concentrate on those types of emergencies that could affect the water quality and therefore potentially put at risk the health of those customers receiving that water.

There is a potential risk of contamination of water at many points from the raw water source to the customer's tap. Water companies and statutory bodies have a whole range of procedures in place to minimise the occurrence of such an incident, and to detail the action necessary to reduce the impact of any incident on the customer. The material that follows can only give an outline of the issues that need to be considered, and the possible courses of action. It should not in any way be taken as a definitive statement of the required action as there are regulatory and legislative instruments in force, which need to be followed using the procedures in place at that time.

9.3.3   Pollution of the raw water

As discussed earlier in the block, and in T210/T237 *Environmental Control and Public Health*, the water abstracted for a potable supply can either be from a surface source or an underground source. The surface sources can have discharges from stormwater overflows, sewage treatment works, manufacturing and industrial plants, mines, animal rearing, and fish farming activities. These are point sources as the discharge is usually from an identifiable pipe into the watercourse. There can also be diffuse sources such as nitrates and pesticides in run-off from land. Underground sources can additionally be affected by saline intrusion near the coast.

In this section, we are concerned with other possible types of pollution such as accidental or deliberate discharges and spillages from farms, industrial sites, and road and rail accidents. Most, if not all, authorised discharges and many accidental spillages are to surface waters. Accidental spillages to underground aquifers are very problematical. While river and stream flows can be identified and temporary dams built to contain pollutants before removal and responsible disposal, the information on underground water flows is often not known with any certainty. Even detecting the presence of a pollutant is not usually possible until it is too late, and then identifying the pollutant plume and removing it is impossible. A possible pollutant into aquifers might be fuel leaking from underground storage tanks and pipework. While there may be some local pollution from the storage system of petrol and diesel from under a garage forecourt, there are large quantities of aviation fuel stored at airports, with greater potential for more serious incidents. There are also trunk pipelines crossing the country, carrying fuel from the refineries to the bulk users and central storage facilities. Another possible source of contamination might be leachate from landfill sites where there is a failure of the liner.

The difficulty in all of these accidental types of pollution is to know that there is a pollution event in the first place. In terms of surface water pollution, if a suitable procedure is set up then it is possible to get information from interested parties. These could include amenity users such as anglers, walkers and joggers, and those who respond to emergencies such as the police, the fire and rescue service, and the industrialists themselves. The Environment Agency has a 24-hour free phone emergency telephone number (0800 807060) for reporting pollution incidents. On receipt of suitable information, it can inform the water company of a possible threat to its abstraction point and appropriate action, such as closing the intake, can be taken. Having closed the river intake and prevented the water entering the storage reservoir or treatment works, the pollutant can be allowed to flow past before the intake is reopened.

Water quality monitors that operate continuously for specific physical and chemical parameters are available. For example, it is common to monitor dissolved oxygen, ammonia, nitrate and turbidity at river water supply abstraction points. Specific pollution monitors, such as the OPAL system, can be put in place. As discussed also, the broad spectrum fish monitor has wide applications for monitoring of pollution. As will be discussed later, incidents involving *Cryptosporidium* are virtually impossible to detect in advance at water treatment works.

Problems at a treatment works can also occur following sudden, exceptional or unexpected changes of raw water quality due to natural causes such as heavy rain. Unless some action is taken, such as altering the coagulant chemical dose, poor quality treated water will be produced.

9.3.4 Emergency at the treatment works

The water leaving a treatment works can be at a consumer's premises in a matter of minutes, or after several days, depending on the particular distribution system. Therefore the water at the final point in the treatment works must be fit for human consumption, as there is little or no opportunity for further remedial treatment or recall of the product.

Waterworks use a considerable amount of electricity for pumping water and for all the ancillary processes associated with water treatment. In case there is an electrical power failure, the works often has a standby generator on-site, or access to a mobile generator. If total power failure occurs, there is no significant health risk to the customer as water will not leave the works without electricity for the final water pumps. Care has to be taken in starting up the works after a shutdown, in order that the treatment processes operate properly with chemicals added correctly, so that all of the water gets satisfactory treatment. For example, with an upflow clarifier the floc blanket will have been completely disrupted by the shut-down and will need to be established again for optimum treatment.

A potentially more serious problem is that of mechanical or electrical failure of, say, a chemical dosing pump. For example, failure of the pump for the coagulant ferric sulphate, or for the disinfectant chlorine solution, could result in a much lower dose than is required, or even no dose at all! Another possible problem is the formation of an air lock in the small pipework of the chemical solutions being dosed, resulting in either no flow or a reduced flow. There should be a whole host of safety features built into the system so that these events are identified very quickly. For example, the pressure loss on a dosing line, or on a proximity switch that responds to flow along pipes, can be used to send out an alarm for certain events. In fact, for events such as the loss of a disinfectant residual, the works will be automatically shut down (as per DWI requirement). The residual disinfectant should be monitored continuously so that no untreated water leaves the treatment works. Often there is a low-level warning alarm and a lower-level shutdown alarm. If, for example, the target free residual chlorine level were 0.5 mg l^{-1}, then a warning (audible and visual) alarm might be generated if the concentration fell to 0.3 mg l^{-1}, and the works could be automatically closed down if the concentration fell below 0.15 mg l^{-1}.

Dosing pumps are often controlled in such a way that they are adjusted automatically to compensate for changes in water flow so that when, for example, the water throughput is increased, more chemical is added proportionally at the same concentration. There may also be feedback from a monitor to adjust the chemical concentration to within narrow limits. Another potential source of accident is if the pump controller suffers a malfunction, causing the chemical to be overdosed. Besides the possibility of having an additional monitor as a watchdog, rather than a controller, the chemical can be dosed from a 'day tank'. This tank is filled from the main storage tank each day, either automatically or manually, but an electrical locking mechanism on the transfer pump prevents the tank from being filled more frequently. If the dosing pump does malfunction, the amount of chemical which can be added is limited, and even if not detected at the works, the dilution within the storage system will help to minimise any effect.

Chlorine is still the most commonly used disinfectant at treatment works. It is often supplied as a liquid under pressure in drums, and in cylinders. While not directly affecting the water supply, a chlorine leak can put operatives and the immediate population at risk from atmospheric contamination. Treatment using ozone is becoming more common but it is generated on-site. Significant ozone leaks to the atmosphere are identified by atmospheric ozone monitors, which can shut off the electricity supply to the ozone generator and provide a fail-safe situation. In both of these events, however, it is essential that no water is put into the supply without having been adequately disinfected.

Other incidents, such as flooding, fire or explosion at the works, are usually of such significance that the event would be identified quickly and the appropriate action taken.

Many water treatment works, and especially the smaller ones, are unattended and only visited by personnel occasionally. In the Camelford incident in north Cornwall in 1988, a delivery of aluminium sulphate was made to an unattended site. If the site was not staffed after normal working hours, the driver was required to contact the control centre, but there was no record of such a contact.

The chemical was put into the chlorine contact tank instead of the chemical storage tank. Over 20 000 customers were affected by the pollution, and this led to considerable adverse publicity for the water company. Following a review by all water companies, a procedure was introduced of having no unattended deliveries, and fitting locks with a special set of keys on valves and flange seals on delivery pipes and tank covers.

Besides failure of dosing equipment, as was mentioned above, changes in raw water quality can require changes to the type and amount of treatment chemicals used. It is essential that at all times the water receives the best treatment under optimum conditions. If the chemicals, such as chlorine, are being added at a rate proportional to water flow and the raw water quality deteriorates so that it has a greater chlorine demand, then there may be no disinfecting properties unless action is taken to adjust the dose.

Cryptosporidium oocysts have been responsible for several outbreaks of disease in this country. There is no simple test that can be carried out to detect the oocysts in raw or treated water on a routine basis. Currently the only certain way the oocysts can be removed from the potable supply is by filtration, as there is a degree of uncertainty regarding the effects of the standard disinfection processes on them. Because of their small size (about 4 to 7 μm diameter) this means that the clarification and filtration processes must be operated efficiently. For example, turbidity can be monitored continuously on the outlet of settlement tanks and individual filter outlets to check on general performance. In case of a value which is outside the required specification, an alarm can be raised and appropriate action can be initiated. For example, a filter could be washed to improve the filtered water supply.

It is important to remember that the oocysts, if present, have been removed by filtration from the treated water but have not been destroyed or inactivated. They will be present in the sludge from the clarification stage and in the backwashings from the filters. It is a common practice for the washings to be put into a settlement tank and, after allowing them to stand, for the supernatant to be returned to the head of the works. Action must be taken at a waterworks implicated in a cryptosporidiosis outbreak to minimise the risk of contamination from the removed oocysts. A possible course of action is to take the settled water from the washwater recovery tanks and put that relatively small volume through a very fine filter to remove the oocysts. The filter can then be treated with steam to inactivate the oocysts before cleaning, as in the Kalsep system described in Section 1.

9.3.5   Problems in the distribution system

You will recall that in T210/T237 *Environmental Control and Public Health* an outline of a distribution system was given. Pipes are laid underground in various patterns or networks to transport water at an adequate pressure and flow from the treatment works to the customer. At one or more points within the network, storage is provided in covered reservoirs to cope with the fluctuating demand, and to provide a reserve for maintenance and other situations. These reservoirs may be elevated in the form of water towers to provide adequate pressure.

There are many potential risks to the water in the distribution system on its way from the treatment works to the customer. It is therefore necessary for constant vigilance to be exercised to prevent serious events from occurring. It is often microbiological contaminants that are considered to be the potential problem in the distribution system. The fire authorities have a legal right to take water from the mains system to fight fires and to check (at fire hydrants) that the water pressure is adequate. This can seriously impact on drinking water quality, and is the subject of continuing discussions between fire authorities and water companies. Infiltration of water through poorly maintained air valves or the ingress of rainwater through flat roofs of storage reservoirs are possible sources of contamination. There is always the threat of vandalism and theft of access covers to the storage reservoirs. Suitable procedures have to be in place, such as intruder alarms and closed-circuit television, to warn of any threat to water quality.

Most water-borne illnesses are associated with viral and protozoan pathogens. However, many of these micro-organisms cannot yet be isolated from water, or detected in routinely used laboratory tests. Consequently, microbiological standards for drinking water safety are set on the basis of the absence of faecal indicator bacteria in a fixed sample volume. That is, the water samples are examined for surrogate indicators of enteric pathogens. Although water leaving a treatment works must be free of faecal contamination, it is not sterile. Some waters, and certainly those that are surface derived, often contain low numbers of non-pathogenic micro-organisms.

Water samples of treated and distributed waters, including those from randomly selected customer taps, are subjected to routine bacteriological tests by the water company supplying the water, for the detection of intestinal bacteria, especially faecal coliforms, and in particular *Escherichia coli*. This is because they are easy to isolate and characterise, and because they are always present in large numbers in the faeces of humans and warm-blooded animals. They are regarded as faecal 'indicators', which by themselves do not usually cause disease, but their presence suggests that pathogens may be present and that the supply is potentially hazardous to health. There is no direct correlation between the numbers of indicator organisms and the actual presence or number of pathogens, or even the risk of illness occurring. The limitations of the microbiological test must be acknowledged. There can be a delay of at least 18 hours before the result is known. Negative tests results may be obtained during an outbreak of say, cryptosporidiosis, because these pathogens can be resistant to levels of disinfectant that eliminate the faecal bacteria indicator organisms.

Ideally, every sample from the distribution system should be free from indicator organisms. It is recognised that this may not be attainable in practice as, for example, the organisms may grow in household taps, although only 'approved materials' should be used. The standards allow for the presence of an occasional total coliform organism but not a faecal coliform organism.

The finding of an indicator organism in a sample of water prompts the need for an on-site investigation to try and identify the cause. It is important to resample immediately at the original location and at nearby properties. It may be that a thorough clean of the inside surfaces of the offending tap will resolve the problem. However, the problem may be more widespread and the pipes may have to be flushed and additional chlorine added at a convenient point, such as at a storage reservoir.

The distribution system should be as watertight as possible to minimise leaks and waste of water. While it is unlikely that any pollutant could enter the system through such an opening due to the water being under pressure, there is a risk of ingress if the pressure is lost due to a burst, or an extra demand on the flow such as for fire-fighting. Loss of pressure can also result in backflow or back-siphonage of contaminated water from incorrectly installed domestic and industrial equipment.

Water companies have the duty to enforce regulations for the prevention of waste, undue consumption, misuse, or contamination of the water supplied by them. For example, in automatic domestic washing machines, the mains water inlet to the machine has an electrically operated solenoid valve followed by an air gap or pipe interrupter, often as a vertical pipe, and a funnel arrangement. This means that it is impossible for dirty water from the washing machine to be drawn back into the water supply as there is no continuous connection between the mains pipe and the washing machine.

There is often a greater risk of contamination from non-standard equipment. For example, there was once an incident at a farm where a concentrated pesticide solution was being diluted in a tank with mains water. The inlet water to the tank should have been from a fixed pipe with the outlet of the pipe above the maximum water level in the tank to provide an air gap and prevent backflow or siphoning. Contrary to the regulations in this instance, the tank was being filled from a hose-pipe. As the tank was taking a long time to fill, the end of the hose was put into the tank and left unattended. The farm was near to the top of a hill. Near the bottom of the hill there was a burst, and the escaping water caused the pesticide-contaminated water to be drawn into the water company's mains. The water was turned off to effect a repair and then turned back on. Although the water in the mains pipe near to the burst was flushed, as is the standard practice, the contaminated water had travelled several miles. Fortunately, the concentrated pesticide solution contained a wetting agent and an indicator dye. Customers receiving the contaminated water noticed that it was highly coloured and frothed, and very quickly alerted the water company, which had the mains pipes flushed and monitored.

Repairs and maintenance to the mains system have to be carried out in a manner similar to a hospital operation. Any excavated hole has to be kept clear of water to prevent ingress of contaminants, and the pipework and components have to be disinfected with the use of chlorine solution. After completing the work, the mains system is flushed locally and a water sample taken for examination from a downstream point. Care has always to be taken when carrying out alterations to the distribution system, so that no cross-connections with any other non-domestic pipework are made.

A simple and economic way of reducing the nitrate concentration in water is by the blending of two or more waters (typically high-nitrate groundwater being diluted to a safe level with river water). However, if there is a pump failure or a burst main leading to loss of blending capacity in the distribution system, then action needs to be taken to protect young children at risk. Bottled water may be distributed to households with infants, for example.

A source of customer annoyance rather than a serious health risk is the disturbance of mains deposits due to a surge along a pipe as a result of a burst, or an incident such as a greater than normal water flow due to use by the fire and rescue service. This can make the water drawn from the tap appear very dirty and particles can block ball valves and stain washing. This event is typical of pressure testing at fire hydrants by fire authority personnel.

9.3.6 Problems within the customer's property

There are many potential problems within a customer's property that could give rise to water quality implications, but they have a limited impact and only affect the occupants. The water companies have a procedure for dealing with customer concerns about water quality, which could include a visit to the property to investigate the problem and to take samples for analysis.

9.3.7 Statutory notifications and action plans

Unlike food and bottled drinks, with mains water there is little or no opportunity to recall the product if it is found to be, or even considered to be, contaminated. It is possible to transmit substances quickly and widely in a community through its water supply. Consequently, it is necessary for health and local government authorities to have in place surveillance and notification mechanisms to identify and limit the impact of any emergency situation.

For over a century, local authorities used to be the providers of water supplies in their area as well as the organisations responsible for its quality. With the formation of the water authorities and, subsequently, the water companies, the local authorities lost the responsibility for supplying water but retained the principle of being involved in the monitoring of water quality for both public and private supplies, through environmental health officers (EHOs). Besides the legal obligations necessary for this monitoring to take place, the public, often for historical reasons, turn to the local authority to express concern about the water supply in the event of dissatisfaction.

For the protection of the health of the community, it is important that all the relevant parties involved act in a sensible and co-ordinated manner in the event of an emergency, such as in a case of water contamination. This applies to the water company, district health authority and local authority, who all have statutory duties to perform. Also, other organisations such as the Environment Agency, the Department for Environment, Food and Rural Affairs, the Department of Health, and the Drinking Water Inspectorate, all need to have a written emergency procedure which should be updated and rehearsed by means of collaborative exercises from time to time.

While water-borne outbreaks are, thankfully, relatively rare, incidents during water treatment and supply are not, but are less likely to be a serious public health hazard. An incident may be identified by routine sampling and analysis, or by plant malfunction. The water company must also have written contingency arrangements to deal with such emergencies, this being a regulatory requirement.

The Consultants in Communicable Disease Control (CCDC) at the health authority have a responsibility for the investigation, management and action following an incident or outbreak of a water-borne disease in the community. The CCDC must have set up and organised the arrangements for an Incident Control Team and an Outbreak Control Team. The Incident Control Team would typically comprise representatives from the CCDC, the Environmental Health Department and the water company. The team could investigate relatively minor incidents, such as isolated cases of contaminated water (say, at the tap of a single household, with nothing found at neighbouring properties). The Outbreak Control Team, on the other hand, is responsible for the investigation and remedial measures associated with major incidents such as an outbreak of cryptosporidiosis. In this larger team, there would be representatives from the CCDC, the Environmental Health Department, the water companies, the local hospital, the Public Health Laboratory, the Community Health Visitors, the Communicable Diseases Surveillance Unit, the Veterinary Surveillance Unit, a representative from the general practitioners of the area, and, possibly, an epidemiologist. The Press Officer from the local hospital could also be an integral part of the team.

Patterns of disease are obtained from medical practitioners, who have a statutory duty to report all cases, or suspected cases, of notifiable disease and of food poisoning as required under the Public Health (Control of Disease) Act 1984 (at the time of writing this block, 2006, cryptosporidiosis and giardiasis are not notifiable illnesses) and the information is provided on a voluntary basis. From the information it has, the CCDC trigger action if there is evidence to suggest that there is an outbreak of a water-borne illness. The CCDC should inform the water company of all instances of cryptosporidiosis in the community to enable the water company to check its systems as soon as possible. It is important to note that the incubation period for cryptosporidiosis is 7–15 days. The Drinking Water Inspectorate (DWI) would be informed of the incident and updated on the progress made in remedying the problem. A report on the whole episode would be submitted to the DWI once the emergency is over. The DWI may undertake its own investigation if it felt that certain aspects were inadequately covered.

In the event of an emergency there are many matters to attend to. Often the initial details are very sketchy and data gathering becomes the first priority.

Once alerted to a possible contamination, the first stage is often to take samples and get them analysed. However, sometimes the water needs screening for groups of chemicals before detailed analytical procedures can be used. Often, in the initial stages it is like looking for a 'needle in the haystack'. Field testing can be very useful for quick results, to get an order of magnitude value, assuming that the contaminant is known and a suitable test kit is available.

Unless there is strong evidence to the contrary, all incidents must presume that the water in the supply system is a potential health hazard and immediate action is required by the water company. Speed is of the essence and so there must be arrangements in place to obtain expert advice on toxicological and microbiological matters from individuals or centres of expertise. Obviously it is necessary to take all reasonable steps to rectify the situation and get supplies back to normal as soon as possible.

There are many courses of action open to the water company in the short term, depending on the number of people affected, and the severity of the situation. The one that is chosen will depend on the circumstances at that time. If one treatment works is affected, then it may be possible to switch to an alternative supply for a short period. In the case of high-nitrate water, it might be possible to provide alternative supplies to particular groups, such as bottled water for babies (as mentioned earlier), if nitrate dilution water is not available. For microbiologically contaminated water, it may be possible to continue to supply, but advise customers not to use the water for drinking and cooking unless it is boiled. The mains water could be turned off completely and water supplied by tanker or bowser, from which the water can be collected by the customers using their own containers. Such water must be boiled before being used for drinking or for food preparation.

In the event of any action that might interrupt supply or alter its quality significantly, there are vulnerable groups, such as home dialysis patients, who would need to be personally informed. Of course, all customers would need to be made aware of the situation and given advice. This could be done by loudhailer vans, leafleting of houses, notices in the press, and through local radio and television stations.

All of the above actions would be carried out with the involvement of the relevant officers of the local authorities and district health authorities according to agreed procedures. It would also be necessary for everyone to keep accurate records of the events and the action taken. Ultimately, following suitable action, the emergency will be at an end and it is then customary to have a 'wash-up' meeting to discuss the event with a view to learning any lessons for the future. The proceedings of such meetings must be recorded.

SAQ 53

Why are faecal coliforms of particular concern if present in a potable water supply, and what action should the water company concerned take?

SAQ 54

(A revision question on water treatment)

A river rises in a rocky mountainous area and flows onto a flat fertile plain which supports a mixed farming economy (dairy and wheat); of late there has been a considerable increase in the intensive rearing of animals. In winter the river has a high flow, while in summer there are often periods of very low flow. In the past, some coal mining took place in the area, and recently there have been problems of mine water rising in the old shafts and entering the river, contaminating the water with iron and manganese in a reduced state. When the metals enter the water, they are oxidised and come out of solution as a red discoloration.

It is intended to use the river as a water source for a nearby town, abstracting the water from the river as it leaves the plain.

What sort of treatment would you expect to be required? Give a flow diagram of the treatment process.

10 WATERWORKS WASTE AND SLUDGES

> **READ**
>
> Set Book Chapter 13: Waterworks wastes and sludges

10.1 Introduction

This chapter considers the wastes and sludges produced at water treatment plants.

The wastes can be classified into the following categories:

- Liquid chemical wastes (such as those produced when ion-exchange resins are regenerated – these are normally discharged to sewer).

- Washings from screens and microstrainers – these are dewatered and disposed of to landfill or incinerated.

- Washwaters from filters and absorbers – these are sent to settlement tanks, and the supernatant is recycled to the head of the works. The sludge generated is sent for treatment.

- Sludge is an inevitable by-product of water treatment, and its treatment and disposal can be costly. Most of the sludge produced at a water treatment works has its origin in one or more of the following:

  suspended solids in the water

  compounds produced when colour removal is effected

  substances precipitated out during the treatment of the water (e.g. iron and manganese)

  coagulants which have precipitated out during treatment

  any solids added during treatment (e.g. powdered activated carbon, bentonite, etc.)

  biological matter (e.g. that removed from slow sand filters during cleaning)

  sludge from water softening.

The amount of sludge produced through using chemicals in water treatment, can be calculated. Rules of thumb also exist (e.g. for sludge produced through colour removal, and in clarifiers and filters). Most sludges are thickened and dewatered, and sent to landfill. Some might be put onto drying beds prior to despatch to landfill. Significant quantities (without being dewatered) are also discharged to sewer. A small proportion of sludges are sent to sludge lagoons (typically 1 m deep) where evaporation can increase the solids content to 10%.

Sludge thickening may be enhanced by the addition of polymer. Dewatering may be by a variety of methods:

- filter plate presses
- centrifuges
- filter belt presses
- vacuum filtration.

Good descriptions of the operation of these systems are given on pp.238–40 of the Set Book.

Sludge produced from the softening of water using the lime-soda process is mineral in nature and easy to dewater. It is, however, thixotropic (i.e. with a high viscosity at low stress, and vice versa, like paint), making it difficult to handle and transport.

10.2   Co-disposal of waterworks sludge with sewage sludge

Waterworks sludge which has been treated can be sent to landfill. In many instances, however, it is possible to transport the sludge in an untreated or partially treated state to the nearest sewage treatment works. Here, the sludge is added to the incoming crude sewage before primary settlement, where it aids settlement and appears as a component of the raw sewage sludge. The sewage sludge can be dewatered, and further treated by anaerobic digestion before disposal.

Some 48% of sewage sludge in the UK is disposed of on farmland. In land disposal, prior anaerobic digestion will have reduced the number of protozoal, viral and bacterial pathogens. Liquid sewage sludges are applied to land by spray irrigation, or injected into the soil, while dried sludges may be spread and then ploughed into the soil. Anaerobically digested sewage sludge typically contains 5% nitrogen, 2% phosphorus and 0.2% potassium. These concentrations are considerably less than the amounts present in commercial fertiliser, but the sludge does have a high organic content, which is valuable, and therefore it can be viewed as a soil conditioner.

10.3   Constituents of sludge important for agriculture

Sewage sludge has certain constituents which are valuable for agriculture. These include:

    organic matter
    plant nutrients
    water content, and
    trace elements.

10.3.1   Organic matter

Untreated sewage sludge contains 70–80% organic matter (by dry weight), most of it derived from body and household wastes. Up to 60% of this (made up of proteins, fats and carbohydrates) is readily biodegradable. The remainder is made up of recalcitrant materials such as ligno-celluloses which degrade slowly.

Organic matter is an essential component of fertile soil, with levels up to 5% (dry weight) being present. The advantage of the organic matter in raw sludge, and its grease content in particular, has been recognised in relation to the binding of light soils and aiding moisture retention during periods of water stress.

10.3.2   Plant nutrients

The major value of sewage sludge is its nitrogen and phosphorus contents, but the other four elements for crop growth (potassium, calcium, magnesium and sulphur) can also be found.

Table 22 shows typical concentrations of nitrogen, phosphorus and potassium in different types of sludges with, for comparative purposes, typical values for farmyard manure and for a general-purpose compound fertiliser. The values are for guidance only, since there may be significant variations in practice.

Nitrogen in untreated liquid sludge is mainly in the form of organic nitrogen which is largely insoluble. Anaerobic digestion converts about 50% of the organic nitrogen to ammoniacal salts which are soluble. Liquid digested sludges may contain 500–1000 mg l^{-1} of ammoniacal nitrogen, representing 1 to 2%

soluble nitrogen on a dry-solids basis. However, if the digested sludge is dewatered, up to 80% of the soluble nitrogen is removed with the water. Undigested sludges, therefore, act as slow-release fertilisers while liquid digested sludges behave as quick-acting nitrogen fertilisers.

Table 22 Typical concentrations of crop nutrients and typical total solids and organic matter contents in various types of sewage sludge with (for comparison) values for farmyard manure and chemical fertiliser

Type of sludge (or other fertiliser)	Total nitrogen	Soluble nitrogen	Phosphorus		Potassium		Total dry solids content[b]	Organic matter content[b]
	(as N)[a]		(as P)	(as P_2O_5) (as % of dry solids)	(as K)	(as K_2O)	(as % of total wet weight)	
Liquid sludge								
Untreated (primary + secondary)	3.5	0.2	1.3	2.9	0.2	0.5	4.5	3.4
Untreated (primary only)	2.5	0.1	1	2.3	0.2	0.5	6	4.2
Untreated (secondary only):								
surplus activated	5	0.2	0.6	1.4	≤0.05	≤0.012	1.5	1.2
humus sludge	4	0.2	0.6	1.4	≤0.05	≤0.012	2.5	1.8
Digested (primary + secondary)	5	2.8	1.8	4.1	0.2	0.5	3.5	2.2
Dewatered sludge (cake)								
Untreated (primary + secondary)	3	0.1	1.1	2.5	0.1	0.25	25	18
Digested (primary + secondary)	3	0.3	1.5	3.4	0.1	0.25	25	15
Other types of fertiliser								
Farmyard manure[c]	1.7	0.5	0.2	0.5	1.3	3.2	35	30
Chemical fertiliser[d]	20	20	4.4	10.1	4.2	10	100	0

Note: In practice, nutrient values will be about ± 30% of the typical values; analyses are normally conducted on individual sludges to enable appropriate rates of application to land to be determined.

[a] As ammoniacal nitrogen or nitrate; [b] total solids, and organic matter content of liquid sludges may be increased by consolidation; [c] nutrient content of farmyard manure is variable depending on origin and maturity; [d] chemical fertilisers are available with a wide range of N:P:K ratios; values given are for a general-purpose fertiliser.

SAQ 55

Why is it desirable to prevent the disposal of both treated and untreated waterworks sludge to a river?

SAQ 56

If a sludge obtained from chemically treated water is discharged to a sewer, what effect could this discharge have on a sewage treatment works?

11   COMPUTER MODELLING IN THE WATER INDUSTRY

11.1   Introduction

Computers are widely used in the water industry. They have become an essential tool and many people in the industry will either have their own personal computer, a networked terminal, or ready access to one or both. Besides the ubiquitous use of computers for writing reports and memoranda, sending and receiving faxes and electronic mail, and for the use of databases and spreadsheets, they are widely used for control, forecasting and modelling. There are also specialist applications of computer packages for tasks such as computer-aided design for construction of new assets, digital mapping of underground assets, and for use in laboratories for the control of sampling programmes and sophisticated analytical equipment, data handling, and reporting of results.

In terms of water management, control can be something simple like increasing or decreasing a process chemical based on some set value, with a suitable monitor and controller providing a feedback loop to the dosing equipment.

Control may also be very sophisticated, as with a river intake protection system where signals from a group of fish are processed and current data on fish behaviour evaluated against historic data (as in the WRc Fish Monitor), to warn of a pollution incident or a significant change in water quality.

In terms of the use of computers for modelling this is usually part of some investigational work, or at some planning stage.

11.2   River modelling

Historically, early models in relation to surface waters concentrated on river flows and hydrological conditions, usually in relation to flood defence or other engineering issues. There was a need for information to manage flood defences, to plan and monitor capital and maintenance programmes, and to issue flood warnings. Catastrophic flood events are thankfully infrequent but this is a disadvantage to information requirements, so modelling is often used for predictions. The initial models were relatively basic, representing steady-state conditions in rivers and using empirical relationships between rainfall, geology and topography to derive surface water run-off characteristics. The next level of development introduced time-varying conditions to understand the effect of flood waters progressing down rivers and the interaction of several hydrographs from the various tributaries in a river system. This was an important refinement as it allowed real-time flood forecasting. The hydrological models were developed to give a more detailed view, allowing the characterisation of subcatchments with differing conditions and including key parameters such as various forms of evaporation, overland flow, interstitial flow, and filtration down to aquifer systems feeding groundwater resources. These were calibrated by specific flood events and then used to predict the outcome of other events, so there was a need to link databases for catchment models and subcatchment models.

The importance of water quality and associated environmental issues has accelerated the expansion of hydraulic models. The early models encompassed parameters such as dissolved oxygen, biochemical oxygen demand of the water, sediment oxygen demand, ammonia and nitrates. More recent developments soluble nitrogen on a dry-solids basis. However, if the digested sludge is dewatered, up to 80% of the soluble nitrogen is removed with the water. Undigested sludges, therefore, act as slow-release fertilisers while liquid digested sludges behave as quick-acting nitrogen fertilisers.

Table 22   Typical concentrations of crop nutrients and typical total solids and organic matter contents in various types of sewage sludge with (for comparison) values for farmyard manure and chemical fertiliser

Type of sludge (or other fertiliser)	Total nitrogen	Soluble nitrogen	Phosphorus		Potassium		Total dry solids content[b]	Organic matter content[b]
	(as N)[a]		(as P)	(as P_2O_5)	(as K)	(as K_2O)	(as % of total wet weight)	
				(as % of dry solids)				
Liquid sludge								
Untreated (primary + secondary)	3.5	0.2	1.3	2.9	0.2	0.5	4.5	3.4
Untreated (primary only)	2.5	0.1	1	2.3	0.2	0.5	6	4.2
Untreated (secondary only):								
surplus activated	5	0.2	0.6	1.4	≤0.05	≤0.012	1.5	1.2
humus sludge	4	0.2	0.6	1.4	≤0.05	≤0.012	2.5	1.8
Digested (primary + secondary)	5	2.8	1.8	4.1	0.2	0.5	3.5	2.2
Dewatered sludge (cake)								
Untreated (primary + secondary)	3	0.1	1.1	2.5	0.1	0.25	25	18
Digested (primary + secondary)	3	0.3	1.5	3.4	0.1	0.25	25	15
Other types of fertiliser								
Farmyard manure[c]	1.7	0.5	0.2	0.5	1.3	3.2	35	30
Chemical fertiliser[d]	20	20	4.4	10.1	4.2	10	100	0

Note: In practice, nutrient values will be about ± 30% of the typical values; analyses are normally conducted on individual sludges to enable appropriate rates of application to land to be determined.

[a] As ammoniacal nitrogen or nitrate; [b] total solids, and organic matter content of liquid sludges may be increased by consolidation; [c] nutrient content of farmyard manure is variable depending on origin and maturity; [d] chemical fertilisers are available with a wide range of N:P:K ratios; values given are for a general-purpose fertiliser.

SAQ 55

Why is it desirable to prevent the disposal of both treated and untreated waterworks sludge to a river?

SAQ 56

If a sludge obtained from chemically treated water is discharged to a sewer, what effect could this discharge have on a sewage treatment works?

---

11 COMPUTER MODELLING IN THE WATER INDUSTRY

11.1 Introduction

Computers are widely used in the water industry. They have become an essential tool and many people in the industry will either have their own personal computer, a networked terminal, or ready access to one or both. Besides the ubiquitous use of computers for writing reports and memoranda, sending and receiving faxes and electronic mail, and for the use of databases and spreadsheets, they are widely used for control, forecasting and modelling. There are also specialist applications of computer packages for tasks such as computer-aided design for construction of new assets, digital mapping of underground assets, and for use in laboratories for the control of sampling programmes and sophisticated analytical equipment, data handling, and reporting of results.

In terms of water management, control can be something simple like increasing or decreasing a process chemical based on some set value, with a suitable monitor and controller providing a feedback loop to the dosing equipment.

Control may also be very sophisticated, as with a river intake protection system where signals from a group of fish are processed and current data on fish behaviour evaluated against historic data (as in the WRc Fish Monitor), to warn of a pollution incident or a significant change in water quality.

In terms of the use of computers for modelling this is usually part of some investigational work, or at some planning stage.

11.2 River modelling

Historically, early models in relation to surface waters concentrated on river flows and hydrological conditions, usually in relation to flood defence or other engineering issues. There was a need for information to manage flood defences, to plan and monitor capital and maintenance programmes, and to issue flood warnings. Catastrophic flood events are thankfully infrequent but this is a disadvantage to information requirements, so modelling is often used for predictions. The initial models were relatively basic, representing steady-state conditions in rivers and using empirical relationships between rainfall, geology and topography to derive surface water run-off characteristics. The next level of development introduced time-varying conditions to understand the effect of flood waters progressing down rivers and the interaction of several hydrographs from the various tributaries in a river system. This was an important refinement as it allowed real-time flood forecasting. The hydrological models were developed to give a more detailed view, allowing the characterisation of subcatchments with differing conditions and including key parameters such as various forms of evaporation, overland flow, interstitial flow, and filtration down to aquifer systems feeding groundwater resources. These were calibrated by specific flood events and then used to predict the outcome of other events, so there was a need to link databases for catchment models and subcatchment models.

The importance of water quality and associated environmental issues has accelerated the expansion of hydraulic models. The early models encompassed parameters such as dissolved oxygen, biochemical oxygen demand of the water, sediment oxygen demand, ammonia and nitrates. More recent developments management. Most aquifers in the UK are closely linked with river systems, to which they contribute baseflow from seepages and springs. As a result, major groundwater abstractions almost inevitably affect river flows.

11.4   Computational fluid dynamics

The rapid development of more powerful computers and dramatic improvements in computational techniques have made computational fluid dynamics (CFD) a very useful tool to both designers and researchers. By solving the interlinked mass, momentum and energy equations, CFD can be applied to simulate complicated two- and three-dimensional flow problems and provide more comprehensive and detailed information than those from experimental measurements on physical models. In CFD models, system parameters such as inlets, outlets or internal baffles can be easily modified to optimise the system performance. CFD can be used to simulate problems in a wide range of flow situations such as for chemical mixing optimisation, and the design of treatment components such as sedimentation tanks, chlorine contact tanks and ozone contact tanks (Figure 20). The flow through a hopper-bottomed upflow settlement tank is very complex and is a good example of a unit that can be modelled to enable desk studies for optimisation to be carried out.

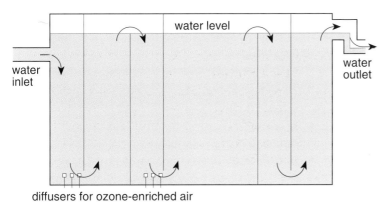

Figure 20   Ozone contact tank

The flow problem is modelled by creating a grid covering the complete area of interest and subdividing the flow domain into a large number of cells. A steady-state solution of the flow domain provides a detailed picture of the flow at each cell. Any short-circuiting or dead space can be identified by studying the velocity vectors. From the steady-state solution, a residence time distribution can be obtained by conducting a transient simulation of a pulse of tracer released at the inlet. The injection of the tracer at the inlet can be followed by measuring the concentration emerging at the outlet. The theoretical residence time of a unit would be the volume of the unit divided by the flowrate. Any short-circuiting would be shown by a residence time shorter than the theoretical value. Any stagnation zones or circulation within the unit will prolong the residence time of part of the flow and therefore this will arrive at the outlet after the theoretical residence time. If the unit being modelled exists, the model can be validated by adding a tracer chemical, such as lithium chloride, to the inlet in the form of a relatively small volume of a concentrated solution, and then samples from the outlet can be analysed to find the physical form of the residence time distribution. The model can be improved, if necessary, by an iterative process until there is good agreement, and then it can be used to evaluate a variety of scenarios to improve the performance. The final solution can then be transferred to the physical unit when the optimum has been found.

11.5   Water distribution system

Dynamic model simulation has been an aid to the design and management of the water distribution network. Network analysis models of the potable water distribution system to give flow and pressure data have been extensively used in the water industry to find out the flow path taken by the water from source to customer, especially for capital investment planning and operational strategies. Driven by the need to improve and maintain customer service levels, the management of water distribution is beginning to rely more on network analysis techniques. The combination of large, detailed models and easy-to-use software is moving network analysis into new areas of benefits. These models have been developed to include techniques for modelling water quality.

Early computer packages only offered basic quality simulation for chlorine decay, pollutant dispersal and blending applications. The situation is rapidly changing and there are now much more complex programmes available. This is because the industry's regulators are now focusing on how water quality varies within a distribution system and the cause and effect of that variation.

Models can now simulate the propagation and concentration of water quality parameters, such as disinfectants or contaminants in dynamic flow conditions. Models have been developed for studies of age of water, substance concentration, and source blending.

The age of the water in the distribution system is important because the longer the water is in the system the greater the chance that water quality will deteriorate. For example, in areas of low flow the mains can get colonised by organisms such as *Asellus* (water hog louse), which if present at a customer's tap can cause consternation – and complaints to the water company!

Chlorine concentration models have been used for the assessment of disinfection performance, chlorine-related taste problems and potential for trihalomethane formation. Another model can simulate the mains debris in the pipework, which has important consequences for management of corrosion of the assets and for customer dissatisfaction resulting from burst pipes, the subsequent loss of supply, and complaints about discoloured water. Hydraulic modelling can be carried out in both the steady-state and dynamic modes. Dynamic models are important due to the very large variations in flow patterns, velocities and source concentrations which take place throughout the day.

To maintain a chlorine residual within the distribution system, it is possible to boost the chlorine levels at strategic points, such as at the outlet of covered treated water service or storage reservoirs. The optimal location and operation of this chlorination equipment to meet disinfection and customer taste requirements can be modelled. It has been found that taste complaints arise when chlorine levels fluctuate, as well as when they are considered too high. Modelling enables various scenarios to be evaluated, including the operation of booster chlorination units within the system, to achieve a more consistent chlorine residual value in the distribution zone with less adverse customer reaction and an improved water quality.

A model could also be used for blending purposes, for example where nitrate concentrations have risen and approached the maximum concentration in drinking water. A dynamic water quality model can be used to manage the supply area, which could have a large number of boreholes, to ensure that the nitrate blend is established to keep the concentration below the maximum allowable value.

If a pollutant enters the distribution system, tracking the movement of this harmful substance through the system is vital. A network model enables this scenario to be carried out under controlled conditions without risk to the customer, and allows the design of possible operational solutions to be evaluated.

11.6   Complete water treatment model

Modelling packages are now available which simulate a complete treatment works. A whole works dynamic model can be built up and used in a variety of control and simulation activities.

First, the physical layout of the works is defined by selecting and linking treatment processes together with the corresponding parameters describing the works operation. At a sewage treatment works this could be, for example, the rate of sludge wastage and concentration of mixed liquor suspended solids. The model can simulate all the different process units and can include nitrification, denitrification and biological phosphorus removal. Secondly, the performance of the works is analysed interactively by adjusting the model parameters and observing the results immediately as the simulation proceeds. At this stage the model is fine tuned to validate it.

Besides being a design tool, the model can be used to look at the plant's response to extreme flows and loads. Units that are stressed and those that have spare capacity can be identified. It is possible to predict the consequences of a vital component (e.g. a pump) breaking down. This allows for rationalisation of standby equipment and the development of effective contingency plans. The simulation with its graphical output can also be successfully used for operator training without the risk of detriment to water quality and the environment.

11.7   River quality modelling

As we have seen above, water quality modelling enables us to make predictions as to the outcome of a set of actions, without actually carrying out a physical simulation. We can represent in mathematical terms the phenomena that we are interested in, and thus study the effect of a change in a given variable on the situation in hand.

In this section the modelling of biochemical oxygen demand, dissolved oxygen, nutrients and total coliforms in a river is considered.

The equations used are introductory and form the basis of water quality modelling. In investigating pollution problems, more sophisticated models considering a host of parameters affecting degradation and dispersion are used.

11.7.1   Biochemical oxygen demand

As explained in T210/T237 *Environmental Control and Public Health*, the biochemical oxygen demand (BOD) of an effluent is the amount of oxygen required for the biodegradation of the pollutants in the effluent by micro-organisms. This is expressed in mg O_2 per litre or g O_2 per m^3. Most often this is the carbonaceous BOD (the biochemical oxygen demand due to the oxidation of carbon-containing compounds) where the effect of nitrifying bacteria (and hence nitrification) is inhibited using allylthiourea (ATU). The equations used to model river quality in the following sections refer to the carbonaceous BOD as this is likely to be much more significant than the oxygen demand due to nitrification. Consider a river (Figure 21) into which a pollutant stream enters at Point A.

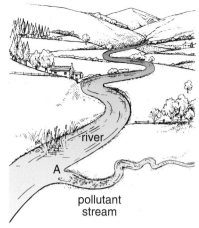

river

A

pollutant
stream

Figure 21   River scenario

If the cross-sectional area of the river is considered constant, under steady-state conditions, the BOD distribution down the river will be given by the expression

$$L = L_0 \exp\left[-\frac{k_\mathrm{L} x}{U_0}\right]$$

(11.1)

where

L is the ultimate carbonaceous BOD or CBOD in the river at a distance x from point A.

L_0 is the ultimate BOD at point A after the pollutant stream enters the river $(\mathrm{g\ m^{-3}})$

k_L is the rate of decay of the BOD in the river $(\mathrm{d^{-1}})$ or BOD decay rate constant

x is the distance from point A (m)

U_0 is the velocity at point A $(\mathrm{m\ d^{-1}})$.

Mass balance can also be used to determine L_0. If

L_up is the ultimate CBOD of the river upstream of point A $(\mathrm{g\ m^{-3}})$

Q_up is the volumetric flowrate of the river upstream of point A $(\mathrm{m^3\ s^{-1}})$

L_e is the ultimate CBOD of the effluent stream $(\mathrm{g\ m^{-3}})$

Q_e is the volumetric flowrate of the effluent stream $(\mathrm{m^3\ s^{-1}})$

then

$$(L_\mathrm{up})(Q_\mathrm{up}) + (L_\mathrm{e})(Q_\mathrm{e}) = L_0(Q_\mathrm{up} + Q_\mathrm{e})$$

$$L_0 = \frac{(L_\mathrm{up})(Q_\mathrm{up}) + (L_\mathrm{e})(Q_\mathrm{e})}{Q_\mathrm{up} + Q_\mathrm{e}}$$

(11.2)

Conversely, 5-day BOD values for the effluent and river can be used to find the 5-day BOD of the mixture and this can be converted to the ultimate BOD by the expression

$$\frac{\mathrm{BOD}_5^{20}}{\left(1 - \mathrm{e}^{-kt}\right)} = L_0$$

(see Example 12 below).

The rate of decay of BOD is temperature dependent, with the relationship between k_L at 20 °C and that at any other temperature T °C for sewage normally being given by

$$k_T = k_{20}(1.047)^{T-20} \qquad (11.3)$$

where

$k_T = k_L$ at T °C

$k_{20} = k_L$ at 20 °C

k_L reflects both the biochemical oxidation of the organic matter and the settling of organic matter on the river bed.

COMPUTER ACTIVITY 2

The solutions to Examples 12 to 16 below have been derived using Excel. You will need to be familiar with the procedure in each of these examples in order to answer the related SAQs.

EXAMPLE 12

A steady flow of partially treated sewage with a BOD_5^{20} of 60 mg l^{-1} enters a river which has a BOD_5^{20} of 2 mg l^{-1}. The temperature of the river ranges from 8 °C to 15 °C over the four seasons of the year.

Given

flowrate of river upstream of discharge point $= 36\ 000$ m^3 h^{-1}

flowrate of treated effluent $= 18\ 000$ m^3 h^{-1}

$k_{20} = 0.25$ d^{-1}

velocity of river $= 0.05$ m s^{-1}

and assuming that the cross-sectional area of the river is constant:

(a) Determine the BOD decay rate constant for each of the temperatures from 8 °C to 15 °C.

(b) Estimate the distance downstream of the discharge point where the ultimate CBOD of the river will be an acceptable 6.3 mg l^{-1} for freshwater fish life for each of the temperatures from 8 °C to 15 °C, and show these values in a graph,

ANSWER

(a) Given that $k_{20} = 0.25$ d^{-1}, we need to calculate k_T for $T = 8$ °C to $T = 15$ °C. Now,

$$k_T = 0.25(1.047)^{T-20} \qquad (11.4)$$

Substituting for relevant values of T in Equation 11.4 gives the following:

T (°C)	k_T (d^{-1})
8	0.144
9	0.151
10	0.158
11	0.165
12	0.173
13	0.181
14	0.190
15	0.199

We can plot this as a graph (Figure 22).

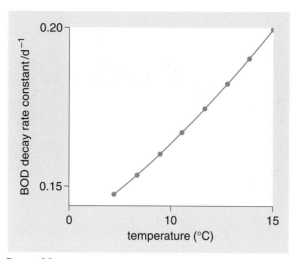

Figure 22

(b) We need to use the equation

$$L = L_0 \exp\left[-\frac{k_L x}{U_0}\right]$$

(11.5)

First, however, we have to calculate the BOD_5^{20} of the effluent/river water mixture and convert it to the ultimate oxygen demand (L_0) of the mixture.

Now, the BOD_5^{20} of the river water is

$$\text{BOD}_{5,\text{up}}^{20} = 2 \text{ mg l}^{-1} \; (= 2 \text{ g m}^{-3})$$

and the BOD_5^{20} of the treated effluent is

$$\text{BOD}_{5,\text{e}}^{20} = 60 \text{ mg l}^{-1} \; (= 60 \text{ g m}^{-3})$$

flowrate of river $(Q_{\text{up}}) = 36\ 000 \text{ m}^3 \text{ h}^{-1} = 10 \text{ m}^3 \text{ s}^{-1}$
flowrate of treated effluent $(Q_{\text{e}}) = 18\ 000 \text{ m}^3\text{h}^{-1} = 5 \text{ m}^3 \text{ s}^{-1}$

BOD_5^{20} of the effluent/river water mixture:

$$= \frac{\left(\text{BOD}_{5,\text{up}}^{20}\right)\left(Q_{\text{up}}\right) + \left(\text{BOD}_{5,\text{e}}^{20}\right)\left(Q_{\text{e}}\right)}{Q_{\text{up}} + Q_{\text{e}}}$$

$$= \frac{(2)(10) + (60)(5)}{10 + 5}$$

$$= \frac{20 + 300}{15} = \frac{320}{15}$$

$$= 21.33 \text{ g m}^{-3} \; \left(= 21.33 \text{ mg l}^{-1}\right)$$

This BOD_5^{20} can be converted to an ultimate oxygen demand by the expression

$$\frac{\text{BOD}_5^{20}}{1 - e^{-kt}} = L_0$$

$$L_0 = \frac{21.33}{1 - e^{-(0.25)(5)}}$$

$$= 29.90 \text{ mg l}^{-1}$$

We thus have $L_0 = 29.90$ mg l^{-1}

Given also $U_0 = 0.05$ m s$^{-1} = 4320$ m d^{-1} and $L = 6.3$ mg l^{-1} and using Equation 11.5, we can calculate the distance x for different values of k_L (covering the temperature range 8–15 °C):

$$L = L_0 \exp\left[-\frac{k_L x}{U_0}\right]$$

$$6.3 = 29.90 \exp\left[-\frac{k_L x}{4320}\right]$$

$$\frac{6.3}{29.9} = \exp\left[-\frac{k_L x}{4320}\right]$$

$$0.211 = \exp\left[-\frac{k_L x}{4320}\right]$$

Taking logs to the base e of both sides

$$-1.556 = -\frac{k_L x}{4320}$$

$$1.556 = \frac{k_L x}{4320}$$

$$x = \frac{(1.556)(4320)}{k_L}$$

$$x = \frac{6721.92}{k_L} \qquad\qquad (11.6)$$

If k_L is fixed, x can be determined.

We have the values of k_L for $T = 8$ °C to $T = 151$ °C. Substituting these into Equation 11.6, we obtain:

T (°C)	k_L / (d^{-1})	x / (km)
8	0.144	46.68
9	0.151	44.52
10	0.158	42.54
11	0.165	40.74
12	0.173	38.86
13	0.181	37.14
14	0.190	35.38
15	0.199	33.78

A plot of x versus T will show the distance downstream of the effluent discharge where the conditions are acceptable for fish life, at different values of the water temperature (Figure 23).

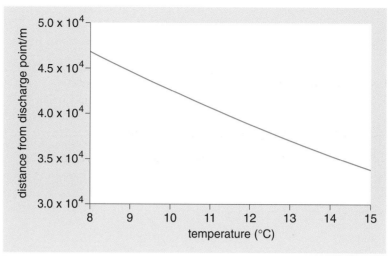

Figure 23

SAQ 57

A water intake is to be constructed downstream of a sewage treatment works (STW) outlet.

(a) If a water treatment plant of category A1 is to be built, calculate how far downstream of the STW outlet the water intake must be, given the following data:

flowrate of river upstream of treated sewage discharge $= 720$ m^3 h^{-1}

BOD_5^{20} of river upstream of treated sewage discharge $= 2$ mg l^{-1}

velocity of river flow $= 0.1$ m s^{-1}

flowrate of treated sewage $= 90$ m^3 h^{-1}

BOD_5^{20} of treated sewage $= 30$ mg l^{-1}

temperature of river water ranges from 6 °C to 14 °C

k_L at 20 °C $= 0.25$ d^{-1}.

(b) Plot a graph relating river water temperature to the positioning of the water intake.

Oxygen demand due to nitrification

Ammonia is oxidised to nitrate in two stages (see T210/T237 *Environmental Control and Public Health*):

$$2NH_4 + 3O_2 \xrightarrow{\textit{Nitrosomonas}} 2NO_2^- + 2H_2O + 4H^+$$

$$2NO_2^- + O_2 \xrightarrow{\textit{Nitrobacter}} 2NO_3^-$$

Based on the above equations, the oxidation of the two atoms of nitrogen in the first stage of nitrification requires six atoms of oxygen, while in the second stage of oxidation two atoms of oxygen are consumed.

In all then, two atoms of nitrogen require eight atoms of oxygen for oxidation.

two atoms of nitrogen $= 2 \times 14 = 28$ atomic mass units

eight atoms of oxygen $= 8 \times 16 = 128$ atomic mass units

$\therefore$ ratio of oxygen demand to nitrogen $= 128/28 = 4.57$

This leads to the following:

nitrogenous biochemical oxygen demand = 4.57 (total oxidisable nitrogen)

where total oxidisable nitrogen = organic nitrogen + ammonia N.

The equations for nitrogen removal are analogous to those for BOD depletion, i.e.

$$N = N_0 \exp\left[-\frac{k_N x}{U_0}\right]$$

where

N is the ultimate nitrogenous biochemical oxygen demand (NBOD) at a distance x from the point of entry of the contaminant stream

N_0 is the ultimate NBOD at the point of mixing between the river and the contaminant stream

k_N is the NBOD decay rate constant, typically 0.3 d^{-1}

U_0 is the velocity of the river (m d^{-1})

N_0 is obtained by mass balance.

Since, in general, the carbonaceous oxygen demand is satisfied first, the modelling of CBOD has taken priority in the analysis of pollution incidents.

11.7.2   Dissolved oxygen

Dissolved oxygen content is frequently used as an indicator of water quality in streams and rivers. The Streeter–Phelps equation given below (Equation 11.7) predicts the dissolved oxygen concentration downstream from a point source of BOD. Assuming a constant river cross-sectional area, the dissolved oxygen deficit D ($= C_s - C$, see T210/T237 *Environmental Control and Public Health*) can be determined by the following expression

$$D = D_0 \exp\left[-\frac{k_a x}{U_0}\right] + \frac{L_0 k_L}{k_a - k_L}\left[\exp\left(\frac{-k_L x}{U_0}\right) - \exp\left(-\frac{k_a x}{U_0}\right)\right] \tag{11.7}$$

where

D is the dissolved oxygen deficit (mg l^{-1}) at a distance x from point of contamination

D_0 is the dissolved oxygen deficit at $x = 0$ (in mg l^{-1})

k_a is the reaeration coefficient (d^{-1}) (see below)

U_0 is the river velocity (m d^{-1})

L_0 is the ultimate CBOD at $x = 0$ (mg l^{-1})

k_L is the BOD decay rate constant (d^{-1}).

Initial dissolved oxygen deficit

The initial dissolved oxygen deficit (D_0) is calculated from:

$$D_0 = C_s - \frac{C_{up}Q_{up} + C_e Q_e}{Q_{up} + Q_e} = \frac{D_{up}Q_{up} + D_e Q_e}{Q_{up} + Q_e}$$

where

D_{up} is the dissolved oxygen deficit upstream of the contamination (mg l^{-1})

Q_{up} is the flowrate of river upstream of the contamination (m^3 s^{-1})

D_e is the dissolved oxygen deficit in the effluent stream (mg l^{-1})

Q_e is the flowrate of the effluent stream (m^3 s^{-1}).

The above equations disregard the oxygen production due to photosynthesis, the oxygen utilisation due to respiration, and the benthic oxygen demand. Photosynthesis is affected by several factors, e.g. turbidity of the water, intensity of sunlight. Similarly, photosynthetic respiration rates vary widely, ranging from 0.5 g m^{-2} d^{-1} to greater than 20 g m^{-2} d^{-1}.

Dissolved oxygen saturation

The saturation concentration of dissolved oxygen (C_s) is a function of temperature (T), salinity and barometric pressure. The effect of salinity becomes important in estuarine systems and, to a lesser extent, in rivers with high irrigation return flows. By far the most important parameter determining the concentration of dissolved oxygen in water is temperature. The temperature dependence of C_s at zero salinity and at sea level is given by the expression:

$$C_s = 14.65 - 0.41022T + 0.00791T^2 - 0.00007774T^3$$

where T is the temperature in °C.

EXAMPLE 13

Given the Streeter–Phelps equation (11.7):

$$D = D_0 \exp\left[-\frac{k_a x}{U_0}\right] + \frac{L_0 k_L}{k_a - k_L}\left[\exp\left(-\frac{k_L x}{U_0}\right) - \exp\left(-\frac{k_a x}{U_0}\right)\right]$$

and the following values

$D_0 = 5$ mg l^{-1}

dissolved oxygen saturation level $= 10.6$ mg l^{-1}

$k_a = 2.5$ d^{-1}

$U_0 = 0.5$ m s^{-1}

$L_0 = 98$ mg l^{-1}

$k_L = 0.25$ d^{-1}

(a) Plot a graph of dissolved oxygen against distance downstream for x from 0 to 600 km in steps of 1.0 km.

(b) From the graph, determine the maximum dissolved oxygen deficit (called the critical deficit, D_c) and the point in the river (the critical distance, x_c) at which it occurs.

ANSWER

(a) We need to find the dissolved oxygen level along the river. Now,

$$DO = DO_{sat} - D$$

$$= 10.6 - \left\{D_0 \exp\left[-\frac{k_a x}{U_0}\right] + \frac{L_0 k_L}{k_a - k_L}\left[\exp\left(-\frac{k_L x}{U_0}\right) - \exp\left(-\frac{k_a x}{U_0}\right)\right]\right\}$$

Substituting for

$$D_0 = 5 \text{ mg l}^{-1}$$
$$k_a = 2.5 \text{ d}^{-1}$$
$$U_0 = 0.5 \text{ m s}^{-1} = 43\ 200 \text{ m d}^{-1}$$
$$L_0 = 98 \text{ mg l}^{-1}$$
$$k_L = 0.25 \text{ d}^{-1}$$

we can generate a plot of dissolved oxygen against distance down the river (Figure 24a).

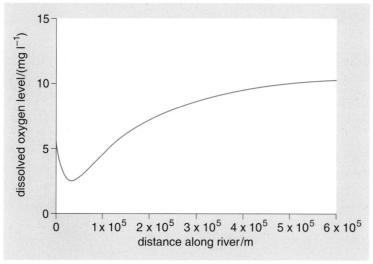

Figure 24(a)

(b) From the graph, the critical (minimum) dissolved oxygen level can be ascertained and its location (the critical distance, x_c) determined (Figure 24b).

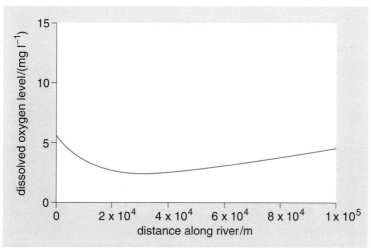

Figure 24(b)

The minimum dissolved oxygen level is about 2.4 mg l^{-1}.

∴ The critical dissolved oxygen deficit = 10.6 − 2.4 = 8.2 mg l^{-1}.

The critical distance is about 33 km from the discharge point.

SAQ 58

An industrial effluent is accidentally discharged into a river.

(a) Given the following values for the pertinent factors in the Streeter–Phelps equation, generate a graph showing the dissolved oxygen concentration for 150 km from the point of the discharge.

$$D_0 = 8.0 \text{ mg l}^{-1}$$

dissolved oxygen saturation level $= 9.8 \text{ mg l}^{-1}$

$$k_a = 2.0 \text{ d}^{-1}$$

$$U_0 = 0.1 \text{ m s}^{-1}$$

$$L_0 = 80 \text{ mg l}^{-1}$$

$$k_L = 0.25 \text{ d}^{-1}$$

(b) Determine D_c and x_c.

Reaeration coefficient

The atmosphere acts as the major source for the replenishment of dissolved oxygen in rivers. There are several correlations for the reaeration coefficient. Three favoured expressions are given below, for a river water temperature of 20 °C. Each expression is valid for a given range of depth and river water velocity (Figure 25).

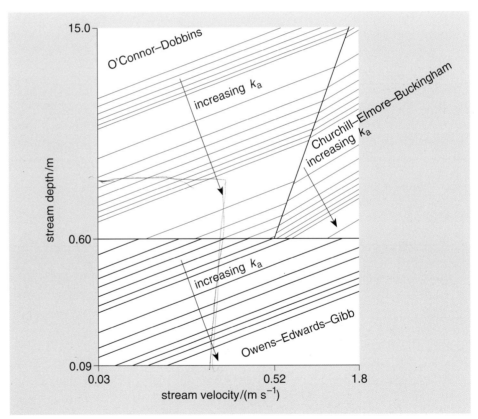

Figure 25   The regions of validity for the different correlations for the reaeration coefficient (k_a)

1 O'Connor–Dobbins correlation

$$(k_a)_{20} = 3.93 \frac{U_0^{0.5}}{H^{1.5}}$$

$$3.93 \times \frac{0.2^{0.5}}{2^{1.5}} = \frac{0.4472135955}{2.828427125}$$

$$\approx 0.621387$$

where

$(k_a)_{20}$ is the reaeration coefficient (d^{-1}) for river water of temperature 20 °C

U_0 is the average stream velocity (m s^{-1})

H is the average stream depth (m)

2 Churchill–Elmore–Buckingham correlation

$$k_a = 5.03 \frac{U_0^{0.969}}{H^{1.673}}$$

3 Owens–Edwards–Gibb correlation

$$k_a = 5.34 \frac{U_0^{0.67}}{H^{1.85}}$$

The reaeration rate is affected by temperature. At a river temperature T °C, the reaeration rate $(k_a)_T$ is approximated by the expression

$$(k_a)_T = (k_a)_{20} (1.024)^{T-20}$$

$$0.621387 (1.024)^{18-20}$$

$$0.6 \, d^{-1}$$

The critical distance and the critical deficit

At the critical distance x_c, the dissolved oxygen deficit will reach a maximum (called the critical deficit, D_c).

The travel time (t_c) to the critical deficit is given by

$$t_c = \frac{1}{k_a - k_L} \ln \left[\frac{k_a}{k_L} \left(1 - D_0 \frac{(k_a - k_L)}{k_L L_0} \right) \right]$$

$$t_c = \frac{1}{\rule{2cm}{0.4pt}}$$

The critical distance can be computed, knowing the travel time and river velocity, i.e.

$$x_c = (U_0)(t_c)$$

The critical deficit (D_c) can be found from

$$D_c = \left[D_0 - \frac{L_0 k_L}{k_a - k_L} \right] \left[\frac{k_a}{k_L} \left(1 - D_0 \frac{(k_a - k_L)}{k_L L_0} \right) \right]^{-\frac{k_a}{k_a - k_L}}$$

$$+ \left[\frac{L_0 k_L}{k_a - k_L} \right] \left[\frac{k_a}{k_L} \left(1 - D_0 \frac{(k_a - k_L)}{k_L L_0} \right) \right]^{-\frac{k_L}{k_a - k_L}} \qquad (11.8)$$

3.6ms¹

EXAMPLE 14

2.0

13000ft

A river of 0.8 m average depth and flowing at 1.5 m s⁻¹ receives a flow of 1.8m⁻¹ treated effluent at 2.0 m³ s⁻¹ from an industrial plant. A mechanical failure in the plant results in an effluent with a BOD_5^{20} of 1500 mg l⁻¹ entering the river. Given the following data, calculate the critical dissolved oxygen deficit and the distance from the effluent discharge point at which it occurs.

> effluent and river temperature = 10 °C
>
> effluent dissolved oxygen concentration = 0 mg l⁻¹   2 mg
>
> k_L = 0.15 d⁻¹ in the river   0.11 e2~
>
> BOD_5^{20} of river upstream of effluent discharge = 3 mg l⁻¹   3.5
>
> river is 80% saturated with oxygen upstream of the effluent discharge
>
> river flowrate = 10 m³ s⁻¹.   flowrate s

Assume the saturation concentration of oxygen in the effluent and in the river is given by the expression:

$$C_s = 14.65 - 0.41022T + 0.00791T^2 - 0.00007774T^3$$

ANSWER

To calculate the critical oxygen deficit, we need to use Equation 11.8.

(a) Determination of D_0:

$$D_0 = \frac{D_{up}Q_{up} + D_eQ_e}{Q_{up} + Q_e}$$

river temperature = 10 °C

saturation concentration of dissolved oxygen (C_s) at 10 °C

$$= 14.65 - 0.41022(10) + 0.00791(10)^2 - 0.00007774(10)^3$$

$$= 11.26 \text{ mg l}^{-1}$$

Since the river water is 80% saturated, the dissolved oxygen concentration in the river water:

$$= 0.80 \times 11.26 = 9.01 \text{ mg l}^{-1}$$

D_{up} = 11.26 − 9.01 = 2.25 mg l⁻¹   deficit

D_e = 11.26 mg l⁻¹   my deficit

Q_e = 2 m³ s⁻¹

Q_{up} = 10 m³ s⁻¹

$$D_0 = \frac{2.25(10) + (11.26)(2)}{10 + 2}$$

$$= 3.75 \text{ mg l}^{-1}$$

(b) Determination of L_0:

$$\text{BOD}_5^{20} \text{ of river water/effluent mixture} = \frac{(3)(10)+(1500)(2)}{10+2}$$
$$= 252.50 \text{ mg l}^{-1}$$

$$L_0 = \frac{\text{BOD}_5^{20}}{1-e^{-kt}}$$

k_L at $10 \text{ °C} = 0.15 \text{ d}^{-1}$

Now, $k_t = k_{20}(1.047)^{T-20}$

$\left(\text{At } T = 10 \text{ °C}, k_{10} = 0.15 \text{ d}^{-1} \right)$

Therefore,

$$k_{20} = \frac{k_T}{(1.047)^{T-20}}$$
$$= \frac{0.15}{1.047^{-10}} = 0.24 \text{ d}^{-1}$$

and

$$L_0 = \frac{252.50}{1-e^{-(0.24)(5)}}$$
$$= 361.33 \text{ mg l}^{-1}$$

(c) Determination of k_a:

Given depth $H = 0.8 \text{ m}$

velocity $U_0 = 1.5 \text{ m s}^{-1}$

Using Figure 23, k_a is best estimated using the Churchill–Elmore–Buckingham correlation:

$$k_a = \frac{5.03 \, U^{0.969}}{H^{1.673}}$$
$$= 5.03 \frac{(1.5)^{0.969}}{(0.8)^{1.673}}$$
$$= 10.82 \text{ d}^{-1}$$

This is for 20 °C. For 10 °C (the temperature in the river),

$k_a = 10.82(1.024)^{10-20} = 8.54 \text{ d}^{-1}$

So, now we have

$D_0 = 3.75 \text{ mg l}^{-1}$

$L_0 = 361.33 \text{ mg l}^{-1}$

$k_L = 0.15 \text{ d}^{-1}$

$k_a = 8.54 \text{ d}^{-1}$.

(d) Determination of D_c:

Substituting in the equation for D_c, the critical dissolved oxygen deficit,

$$D_c = \left[D_0 - \frac{L_0 k_L}{k_a - k_L} \right] \left[\frac{k_a}{k_L} \left(1 - D_0 \frac{(k_a - k_L)}{k_L L_0} \right) \right]^{-\frac{k_a}{k_a - k_L}}$$

$$+ \left[\frac{L_0 k_L}{k_a - k_L} \right] \left[\frac{k_a}{k_L} \left(1 - D_0 \frac{(k_a - k_L)}{k_L L_0} \right) \right]^{-\frac{k_L}{k_a - k_L}}$$

$$= \left[3.76 - \frac{(361.33)(0.15)}{(8.54 - 0.15)} \right] \left[\frac{8.54}{0.15} \left(1 - 3.76 \frac{(8.54 - 0.15)}{(0.15)(361.33)} \right) \right]^{-\frac{8.54}{8.54 - 0.15}}$$

$$+ \left[\frac{(361.33)(0.15)}{(8.54 - 0.15)} \right] \left[\frac{8.54}{0.15} \left(1 - 3.76 \frac{(8.54 - 0.15)}{(0.15)(361.33)} \right) \right]^{-\frac{0.15}{8.54 - 0.15}}$$

$$= 6.03 \text{ mg l}^{-1}$$

(e) Determination of x_c:

The distance at which the critical dissolved oxygen deficit occurs

$$x_c = U_0 t_c$$

$$U_0 = 1.5 \text{ m s}^{-1} = 129\ 600 \text{ m d}^{-1}$$

$$t_c = \frac{1}{k_a - k_L} \ln \left[\frac{k_a}{k_L} \left(1 - D_0 \frac{k_a - k_L}{k_L L_0} \right) \right]$$

Now

$$k_a = 8.54 \text{ d}^{-1}$$
$$k_L = 0.15 \text{ d}^{-1}$$
$$D_0 = 3.75 \text{ mg l}^{-1}$$
$$L_0 = 361.33 \text{ mg l}^{-1}.$$

Substituting in the equation,

$$t_c = \frac{1}{8.54 - 0.15} \ln \left[\frac{8.54}{0.15} \left(1 - 3.75 \frac{(8.54 - 0.15)}{(0.15)(361.33)} \right) \right]$$

$$= 0.38$$

$$\therefore x_c = (129\ 600)(0.38) = 49\ 248 \text{ m} = 49.25 \text{ km}$$

From the above, the critical dissolved oxygen deficit is 6.03 mg l^{-1}. This equates to a dissolved oxygen level of $11.26 - 6.03 = 5.23$ mg l^{-1}.

The critical deficit occurs 49.25 km downstream of the effluent discharge point.

SAQ 59

An illegal discharge of farmyard slurry with a BOD_5^{20} of 450 mg l^{-1} enters a river. Given the information below, determine the critical dissolved oxygen deficit and its location with reference to the entry point of the slurry into the river.

average river depth $= 0.7$ m

river flowrate $= 2.0$ m^{3}s^{-1}

average river velocity $= 0.2$ m s^{-1}

BOD_5^{20} of river upstream of farm $= 1.5$ mg l^{-1}

flowrate of slurry $= 0.4$ m^3 s^{-1}

k_L at 20 °C $= 0.3$ d^{-1}

dissolved oxygen concentration in river upstream of effluent discharge $= 7$ mg l^{-1}

Assume the saturation concentration of dissolved oxygen in the river and the effluent is that of clean water.

The effluent and river water are both at 20 °C and the effluent has a dissolved oxygen content of 0.1 mg l^{-1}.

Effect of nitrification

Nitrification can be significant in rivers where nitrifying bacteria have become established due to, say, N-bearing pollutant streams discharging into the river and where a stable river substrate facilitates their attachment and colonisation.

If nitrogenous BOD is to be taken into account, the expression for the dissolved oxygen deficit becomes:

$$D = D_0 \exp\left[-\frac{k_a x}{U_0}\right] + \frac{L_0 k_L}{k_a - k_L}\left[\exp\left(-\frac{k_L x}{U_0}\right) - \exp\left(-\frac{k_a x}{U_0}\right)\right]$$

$$+ \frac{N_0 k_N}{k_a - k_N}\left[\exp\left(-\frac{k_N x}{U_0}\right) - \exp\left(-\frac{k_a x}{U_0}\right)\right]$$

where

N_0 is the ultimate nitrogenous BOD (NBOD) in the river after entry of the pollutant stream

k_N is the decay rate of NBOD in the river.

Nitrification is very much pH dependent, with an optimum pH range of 8.0–8.5. It is unlikely to be significant below pH 7.0.

The temperature dependence of k_N is the same as that of k_L, i.e.

k_N at temperature $T = (k_N)_{20} (1.047)^{T-20}$

11.7.3 Nutrients and eutrophication potential

The presence of excess plant nutrients can lead to eutrophication. A possible representation for the stoichiometry of algal growth is as follows:

$$106CO_2 + 16NO_3^- + HPO_4^{2-} + 122H_2O + 18H^+ + \text{trace elements} + \text{energy}$$

$$\underset{R}{\overset{P}{\rightleftarrows}} \quad [C_{106}H_{263}O_{110}N_{16}P_1] + 138O_2$$

algal protoplasm

where

P is the process of photosynthesis

R is the process of respiration.

The ratio of C:N:P in the algae in terms of atomic mass can be seen to be 106:16:1. Converted to weights, this comes to 41:7.2:1. Thus only a small amount of phosphorus is needed, in relation to the amounts of carbon and nitrogen, to support algal growth. If the phosphorus is not present in the amount required, algal production will be curtailed, regardless of how much of the other nutrients is available. In this situation, phosphorus is termed the growth limiting element.

Nitrogen uptake by algae is generally in the form of nitrate, if available. Different types of freshwater algae, however, can utilise organic nitrogen, or inorganic nitrogen in the form of ammonia. Algal uptake of phosphorus is usually in its inorganic form as the orthophosphate ion.

Plotting the TN:TP ratio can give an indication as to which element is limiting growth of the algae. Conversely, it can give information on what control measures to take to prevent eutrophication.

It is assumed that other factors such as temperature and river velocity are favourable to algal growth in the given watercourse.

Estimating in-stream nutrient concentrations

If total-nitrogen (TN) and total-phosphorus (TP) can be considered conservative (i.e. not reactive and remain either in solution or suspension), a mass balance approach can be formulated for these constituents.

(In reality, however, this assumption may not be met for a variety of reasons, e.g. algae utilise nutrients, die, and settle to the bottom. Assuming total-N and total-P to be conservative, however, should give an estimate of the upper limit of the in-stream concentrations of these elements.)

The in-stream concentration of total-N (TN_0) resulting from a point discharge (e.g. an effluent stream) is given by:

$$TN_0 = \frac{TN_{up}Q_{up} + TN_eQ_e}{Q_{up} + Q_e}$$

where

TN_0 is the resulting in-stream TN concentration (mg N l^{-1})

TN_{up} is the in-stream TN concentration upstream of the discharge (mg N l^{-1})

Q_{up} is the flowrate of the river upstream of the point of discharge (m^3 s^{-1})

TN_e is the concentration of TN in the point discharge (mg N l^{-1})

Q_e is the flowrate of the point discharge (m^3 s^{-1}).

For a diffuse or distributed discharge (say run-off from a field of crops) the in-stream concentration of TN would be given by:

$$TN = TN_0 + \frac{\Delta Qx}{Q}(TN_r - TN_0)$$

(this assumes no removal of N by bacterial or other action)

where

TN_0 is the in-stream TN concentration at $x = 0$ (mg N l^{-1})

ΔQ is the incremental flow increase per unit distance (m³ s⁻¹ km⁻¹)

x is the distance downstream from the reference point (km)

Q is the stream flowrate at x (m³ s⁻¹)

TN_r is the concentration of TN entering with the distributed flow (mg N l^{-1}).

Equations for total-P are analogous to those for total-N, i.e. for a point source of P:

$$TP_0 = \frac{TP_{up}Q_{up} + TP_eQ_e}{Q_e + Q_e}$$

and for a diffuse discharge of P:

$$TP = TP_0 + \frac{\Delta Qx}{Q}(TP_r - TP_0)$$

EXAMPLE 15

Inappropriate application rates of chemical fertiliser result in N and P being washed into a river flowing for 10 km across an agricultural area of the country. Given the details below:

(a) determine the concentration profiles of total-N and total-P along the length of the river through the farmland;

(b) plot the ratio TN:TP as a function of distance.

  initial flowrate of river (i.e. without contribution of run-off) = 20 m³ s⁻¹

  flowrate of run-off over the 10 km length of river = 4 m³ s⁻¹

  initial concentration of total-N in river = 0.05 mg l^{-1}

  initial concentration of total-P in river = 0.02 mg l^{-1}

  concentration of total-N in run-off = 1.0 mg l^{-1}

  concentration of total-P in run-off = 0.25 mg l^{-1}

ANSWER

(a) The concentration profile for total-N in the river is given by the equation

$$TN = TN_0 + \frac{\Delta Qx}{Q}(TN_r - TN_0) \tag{11.9}$$

Given that

$TN_0 = 0.05$ mg l^{-1}

$\Delta Q = 4/10 = 0.4$ m³ s⁻¹ km⁻¹

Q is the flowrate at $x = 20 + 0.4x$ m³ s⁻¹

$TN_r = 1.0$ mg l^{-1}

Substituting in Equation 11.9,

$$TN = 0.05 + \frac{0.4x}{20 + 0.4x}(1.0 - 0.05)$$

$$= 0.05 + \frac{0.4x}{20 + 0.4x}(0.95)$$

$$= 0.05 + \frac{0.38x}{20 + 0.4x}$$

We can plot a graph of TN versus distance for the 10 km stretch of river. This is shown in Figure 26.

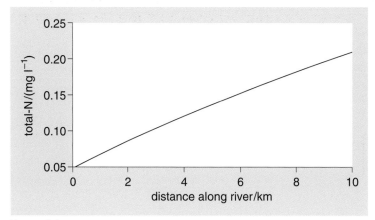

Figure 26

The concentration of total-P in the river can be calculated using the equation

$$TP = TP_0 + \frac{\Delta Qx}{Q}(TP_r - TP_0)$$ (11.10)

Given

$TP_0 = 0.02$ mg l^{-1}

$\Delta Q = 0.4$ m^3 s^{-1} km^{-1}

$Q = 20 + 0.4x$ m^3 s^{-1}

$TP_r = 0.25$ mg l^{-1}

Substituting in Equation 11.10,

$$TP = 0.02 + \frac{0.4x}{20 + 0.4x}(0.25 - 0.02)$$

$$= 0.02 + \frac{0.4x}{20 + 0.4x}(0.23)$$

$$= 0.02 + \frac{0.092x}{20 + 0.4x}$$

As in the case for total-N, we can plot a graph of total-P versus distance (Figure 27).

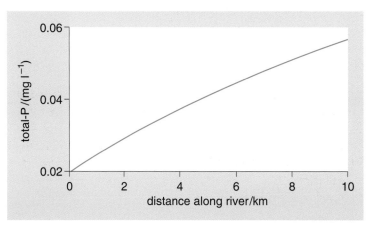

Figure 27

(b) A plot of the ratio TN:TP for points along the river can be generated. From this graph, the TN:TP ratio at a given point can be ascertained. (Conversely, the location of a given TN:TP ratio can be found. For instance, the TN:TP ratio of 3.0:1 occurs at a distance of about 2 km from the start point ($x = 0$) (Figure 28).)

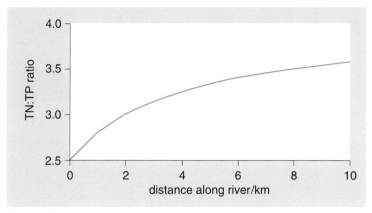

Figure 28

SAQ 60

A 15 km stretch of river flows across agricultural land on which wheat is grown. Nitrate and phosphate from artificial fertilisers are leached out of the fields and enter the river. Given the data below,

(a) determine the concentration profiles of total-N and total-P along the length of the river;

(b) determine the point in the river where the limiting element changes from one plant nutrient to the other, assuming the native species of algae has an N:P ratio of 7.5:1.

Assume that the input of run-off and the concentration of the compounds in it are uniform along the length of the river, and that all other elements (apart from N and P) needed for algal growth are present in excess.

flowrate of river upstream of wheatfields $= 10$ m^3 s^{-1}

flowrate of run-off along the 15 km stretch of river $= 5.0$ m^3 s^{-1}

concentration of total-N in river upstream of wheatfields $= 0.3$ mg N l^{-1}

concentration of total-P in river upstream of wheatfields $= 0.03$ mg P l^{-1}

concentration of total-N in run-off $= 1.5$ mg l^{-1}

concentration of total-P in run-off $= 0.25$ mg l^{-1}.

11.7.4　Total coliform bacteria

Coliform bacteria are considered an indicator of the presence of pathogenic organisms and, as such, relate to the potential for public health problems. The allowable levels of total coliform bacteria in a waterway vary with the intended use of the water: for example, in water for bathing, in a river or in the sea, the EU Bathing Water Directive (76/160/EEC) sets a guideline limit of 500 per 100 ml for this parameter.

A major source of coliforms is treated sewage but the contribution from urban stormwater run-off can also be significant, especially through combined sewer overflows.

For a point source of coliforms, such as a partially treated effluent, the equation representing total coliforms is:

$$TC = TC_0 \exp\left[-\frac{k_{tc}x}{U_0}\right]$$

where

TC is the total coliform number (MPN/100 ml) at a point x m from the point of discharge

TC_0 is the initial total coliform number (MPN/100 ml) (see below)

k_{tc} is the decay coefficient for total coliforms (d^{-1})

x is the distance downstream from point of discharge (m)

U_0 is the velocity of the river (m d^{-1})

The initial total coliform number is

$$TC_0 = \frac{(TC_{up})(Q_{up}) + (TC_e)(Q_e)}{Q_{up} + Q_e}$$

where

TC_{up} is the total coliform number upstream of the discharge point (MPN/100 ml)

Q_{up} is the flowrate of the stream upstream of the discharge point (m^3 s^{-1})

TC_e is the total coliform number in the effluent stream (MPN/100 ml)

Q_e is the flowrate of the effluent (m^3 s^{-1})

The decay coefficient varies with temperature in the following manner:

$$k_{tc} = 1.0 + 0.02(T - 20)$$

where T is the water temperature in °C.

EXAMPLE 16

A sewage treatment works discharges a poor effluent with a total coliform level of 4×10^6 per 100 ml. If the effluent is discharged into a river with a background total coliform count of 800 per 100 ml, given the data below,

(a) estimate the distance downstream from the sewage treatment works outfall where it would be safe to swim;

(b) plot a graph showing the decrease in the total coliform number with distance down the river.

river flowrate = 20 m^3 s^{-1}

effluent flowrate = 5 m^3 s^{-1}

river and effluent temperature = 10 °C

river velocity = 0.15 m s^{-1}.

12   EMERGENCY WATER SUPPLY SYSTEMS

12.1   Introduction

So far we have considered how water is treated for potable supply in circumstances where equipment and funding are not major constraints. Increasingly, there are situations where both the availability of equipment and funding present challenges to the water engineer. Typical of these is the case where water supply is required for large numbers of displaced people in refugee camps.

Surface and ground waters are exploited to provide the bare minimum of 15–20 litres per head per day. Various charities, such as Oxfam and Médecins Sans Frontières, often airfreight in the necessary water treatment and supply equipment, and volunteer engineers recruited through organisations such as RedR (Registered Engineers for Disaster Relief) lend their expertise in keeping the systems operational.

12.2   A possible treatment system

A typical Oxfam water treatment and supply system comprises pumps, storage tanks, a slow sand filter, a disinfection unit, contact tank, distribution pipes and standpipes. All the items are made for easy transport and rapid assembly on site. Figure 31 shows a typical scheme.

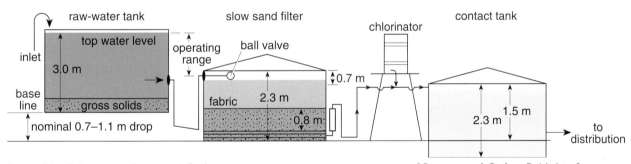

Figure 31   Schematic of a typical Oxfam emergency water treatment system [Courtesy of Oxfam Publishing]

Pumps are used to draw underground or surface water into a large storage tank (45, 70 or 95 m³ in volume). Ideally, underground water would be used, as this would be less polluted. If surface water is to be exploited, it can be abstracted from sediments in the river bed. Two methods are possible:

(a) By construction of a well in the riverbank which is surrounded by gravel and connected to the porous river sediments by a sand-filled channel (Figure 32).

(b) By use of a slotted or perforated pipe laid under the riverbed. The pipe is surrounded by graded gravel and sand. (Note: gravel and sand backfill should be free of iron, otherwise the treatment process will be complicated.) It leads water to a collection chamber in the riverbank, from which it is pumped to the storage tank. The yield may be increased by construction of a subsurface dam, a wall across the riverbed, which blocks the subsurface flow, downstream from the intake pipe. Such a dam will also protect the pipe from movement (Figure 33).

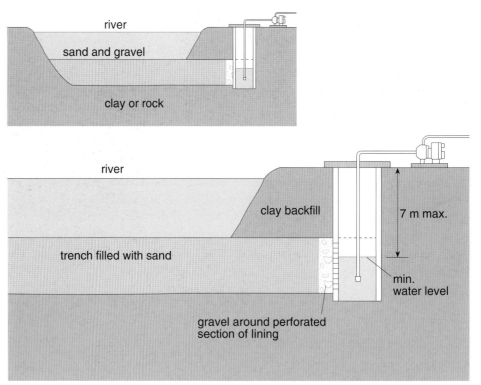

Figure 32   Construction of a well in the riverbank [Courtesy of Oxfam Publishing]

In the storage tank, sedimentation takes place. The tank is usually made of corrugated steel sheets, with a reinforced synthetic rubber liner and a PVC cover. Chemical coagulants may be used to speed up settlement of solids. In certain parts of the world, natural coagulants may be available, saving on the costs and difficulties of using chemicals. For instance, the kernels from the seeds of the *Moringa* plant contain water-soluble proteins which possess a net positive charge, and act as a coagulant.

After settlement the water flows by gravity to one of two sand filters (duty and standby), which consists of a variety of components (Figure 34). A 100 mm layer of sand is spread on the base of the tank on which a network of perforated UPVC piping is laid to collect the filtered water. The pipe is covered with a 125 mm layer of gravel. Above this is placed a layer of geotextile fabric (2 mm thick) which allows water but not sand to pass through. Above the fabric is placed 800 mm of sand. Four layers of geotextile fabric are then placed over the sand and weighted down with stones. The *Schmutzdecke* is established on the geotextile fabric, rather than on the sand as it is on a conventional slow sand filter. The water depth during operation of the filter is maintained at a depth of approximately 1m above the sand.

When the filter flow becomes unacceptably low, cleaning has to be undertaken. Since cleaning and re-establishment of the *Schmutzdecke* takes some three days, the standby filter has to be brought on-stream to maintain the water supply. For optimum effectiveness, the standby filter should be brought on-stream three to four days before use, to help establish the *Schmutzdecke*. In cleaning, the geotextile membranes on top of the sand are carefully removed and thoroughly hosed down to remove all the silt and organic material. This is a faster and more sustainable means of managing the filter, compared with what happens in a conventional slow sand filter.

The membranes are then replaced and water is passed through the filter until the *Schmutzdecke* is re-established. During this phase, the output from the filter is allowed to flow to waste. Once the quality of the output is satisfactory, the water is again directed into the distribution system.

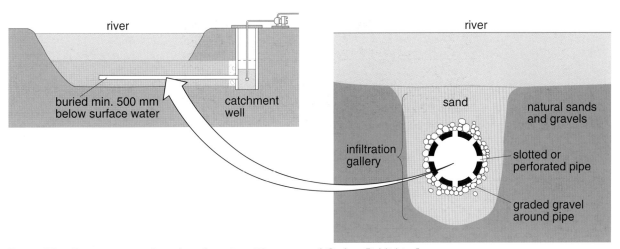

Figure 33   Construction of a subsurface dam [Courtesy of Oxfam Publishing]

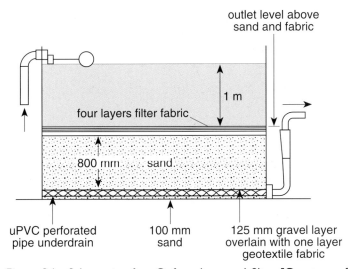

Figure 34   Schematic of an Oxfam slow sand filter [Courtesy of Oxfam Publishing]

From the slow sand filter the water flows into a chlorine contact tank, at the entrance of which a simple floating bowl chlorinator in a clean oil drum, suitably coated or lined to prevent corrosion (Figure 35), doses the water with sodium hypochlorite solution at a constant rate. If the hypochlorite were dosed straight from the drum, the flowrate would initially be high (due to the hydrostatic head of the drum contents) and then would diminish as the hypochlorite level decreased.

The floating bowl chlorinator (Figure 36) consists of a plastic bowl with a cork at the bottom of the bowl; the cork has three glass, brass or copper tubes passing through it. The bowl is placed on the surface of a drum of sodium hypochlorite solution. A flexible hose is connected to one of the tubes (A) on the cork and runs to the outlet of the drum. Through one of the other two tubes (B), a nylon string is threaded, and the string is made taut (by tying a brick to the lower end, and tying the upper end to the lid of the drum). This stops the bowl catching on the sides of the drum as the sodium hypochlorite is used up.

As the liquid level in the drum falls, the bowl moves down with it, always floating on the surface. Stones are put in the bowl to make it float straight and steady. The flow from the chlorinator can be adjusted by carefully positioning tube (C), which lets the sodium hypochlorite solution into the bowl. The flow is reduced by moving it upwards to reduce the height 'H' between its tip and the liquid level in the tank. The flow from the chlorinator is stopped by lifting the bowl out of the solution. (The outlet should not be shut off or the bowl will fill

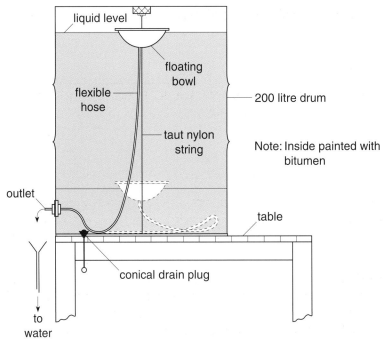

Figure 35　Floating bowl chlorinator, to feed chlorine solution at a constant rate [Courtesy of IRC International Water and Sanitation Centre]

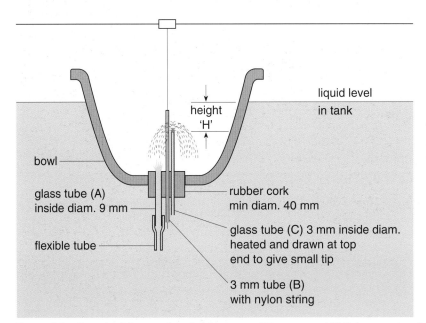

Figure 36　Detail of floating bowl chlorinator [Courtesy of IRC International Water and Sanitation Centre]

up and sink to the bottom of the tank. Accidental sinking can, however, be avoided by fitting a float to the bowl.) It is necessary to stop chlorinating when the pumps are not sending water through the system, for instance.

The drum used for the sodium hypochlorite solution must be coated with bitumen to prevent attack by chlorine. The bitumen must be properly cured prior to use, as otherwise chloro-bituminous tastes will be produced in the treated water. Dissolved bitumens may also produce unacceptable hydrocarbon levels. The tank should have a cover over the top to keep out light (which would help break down the sodium hypochlorite solution).

From the chlorine contact tank, the water flows by gravity through PVC pipes to standpipes where people can collect the water in buckets or other containers for their use. The taps should preferably be of a water-saving variety. One type closes itself when released; another only delivers a set amount of water when operated.

SAQ 62

Describe the main components of an emergency water treatment system that could be set up easily and relatively cheaply.

12.3   Portable water purification equipment

Compact, portable water purification equipment for emergency use is usually readily available from specialist suppliers. These are often sophisticated systems able to treat fresh, brackish, saline and even chemically contaminated waters at a very high rate (up to 23 m^3 h^{-1}).

One such system (Figure 37) comprises a pump set, a filter unit and a steriliser module. The pumps draw raw water and pass it through plastic or stainless steel filter elements (Figure 38a), which have a precoat of fine filter media (such as diatomaceous earth). Suspended solids (including viruses, bacteria and cysts) are removed by the filter.

Figure 37   A portable water purification system [Courtesy of Stella-Meta, part of ITT Industries, Inc]

When blocked, the filter is cleaned using compressed air (rather than by using water, as in sand filters). Figure 38(b–d) shows the sequence of operations for cleaning.

The filtered water is sterilised using sodium hypochlorite solution. If the transport of sodium hypochlorite is to be avoided (say, for safety reasons) it can be produced on-site by the electrolysis of brine. Electrolysis of 3% NaCl brine produces a 1% solution of sodium hypochlorite solution suitable for dosing.

For purification of brackish or saline waters, reverse osmosis membranes are used to lower the total dissolved solids content of the water. In the case of

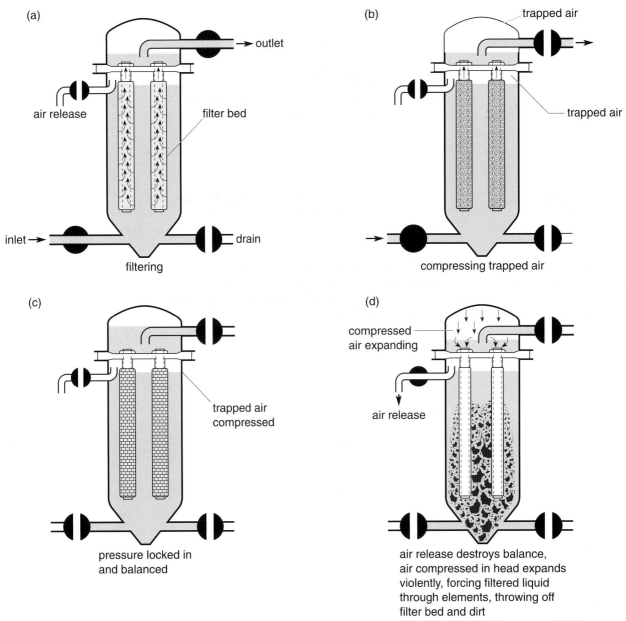

Figure 38 The filtration module (a) and its cleaning sequence (b) to (d) [Courtesy of Stella-Meta, part of ITT Industries, Inc]

chemical contaminants being present in the water, activated carbon filters are used in the final stage to adsorb the impurities.

In addition to being used in emergency or disaster solutions, portable water purification systems are widely used by armed forces personnel to produce large quantities of water rapidly in the field.

12.3.1 Water 'sterilisation' tablets

Water 'sterilisation' tablets offer a convenient means of disinfecting small volumes of water, such as for personal consumption. In one brand of such tablets (Aquatab™), chlorine is present in the form of the organic compound sodium dichloroisocyanurate. The tablets are available in various dosages (8.5 mg to 8.7 g) to cater for different volumes of water to be treated.

Sodium dichloroisocyanurate ($C_3Cl_2N_3NaO_3$) is a white powder with a molecular mass of 219.9. On dispersal in water, hypochlorous acid (the active disinfecting agent) and monosodium cyanurate (a non-toxic compound) are quickly liberated (Figure 39). It has been used in the past to disinfect swimming pools and spa baths.

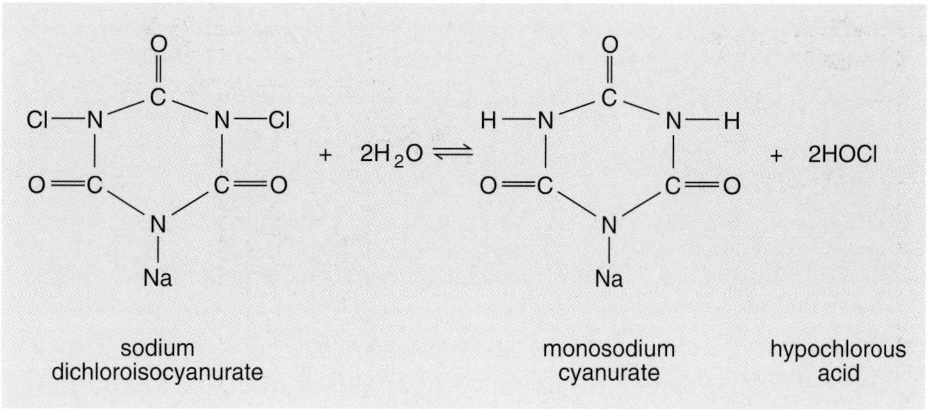

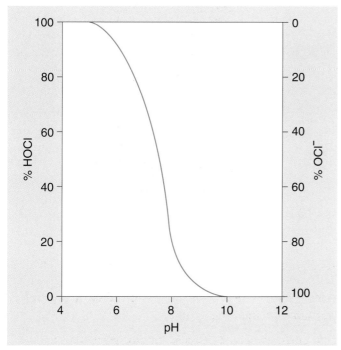

Figure 39

The biocidal capacity of sodium dichloroisocyanurate is claimed to be greater than that of the commonly used disinfecting chemicals such as sodium hypochlorite or calcium hypochlorite, due principally to the lower pH generated in the water when it is used. A low pH results in a higher percentage of undissociated HOCl in the water (Figure 40).

Figure 40   Relative amounts of HOCl and OCl⁻ in water at 20 °C [From Sundstrom and Klei, 1979]

When using the water 'sterilisation' tablets, it is recommended that a free chlorine residual of $\geqslant 0.5$ mg l^{-1} be achieved, following a minimum 30 minutes contact time (as per WHO Guidelines), at a pH of less than 8, in water with a turbidity less than 1 NTU.

Where it is not possible to undertake the above measurements due to, say, lack of facilities or trained personnel or in emergency situations, the guidelines in Table 23 may be appropriate. In all instances, there should be a faint odour of chlorine 30 minutes after addition of the tablet(s).

Table 23   Guidelines for adding chlorine

Water source and mode of supply	Free available chlorine to be added (mg l^{-1})
Piped water supply, house connection; where the water is clear.	1
Protected standpipe, tubewells with pump, covered/protected wells, covered rainwater collection systems; where the water is clear.	2
Unprotected wells, open dug wells; where the water may be cloudy (turbid). Any sediment must be allowed to settle or the water filtered through a fine cloth. The decanted or filtered water is treated.	2–5
Streams, rivers or ponds. Any sediment must be allowed to settle or the water filtered through a fine cloth. The decanted or filtered water is treated.	2–5
Water known to be faecally contaminated. Any sediment must be allowed to settle or the water filtered through a fine cloth. The decanted or filtered water is treated.	5–10

SAQ 63

Which of the following statements are false?

A   In compact, portable water purification units, a filter similar to a slow sand filter is used.

B   Diatomaceous earth filters can remove bacteria.

C   In the cleaning of diatomaceous earth filters, both compressed air and water are used.

D   Using electrolysis, sodium hypochlorite can be produced from brine.

E   At about pH 5, the proportion of hypochlorite ions is at its maximum in water at 20 °C.

F   With water 'sterilisation' tablets, a free chlorine residual of $\geqslant 0.5$ mg l^{-1} is recommended after a minimum contact time of 30 minutes, at a pH of less than 8, with water of turbidity $\leqslant 1$ NTU.

12.4   Emergency sanitation

Equally as important as emergency water supply is emergency sanitation, as a breakdown in the latter can lead to disease and loss of life through the contamination of ground and surface waters. Emergency sanitation covers a number of areas, as you can imagine, such as excreta disposal, wastewater management, solid waste management, waste management at medical centres, disposal of dead bodies, and hygiene promotion. We will consider the water and sewage aspects in this block.

12.4.1   The pit latrine

The most common technique adopted for excreta disposal in emergencies is the pit latrine (Figure 41). It is simple, relatively easy to construct, inexpensive, and operates without water. The pit should be at least 2 m in depth, and covered by a latrine slab, with a drop-hole to allow excreta to drop straight into the pit. The drop-hole can be covered with a removable lid to minimise fly and odour problems. The slab should be raised above the surrounding ground, to prevent surface water entering the pit. If the soil is unstable, the pit should be lined to prevent collapse.

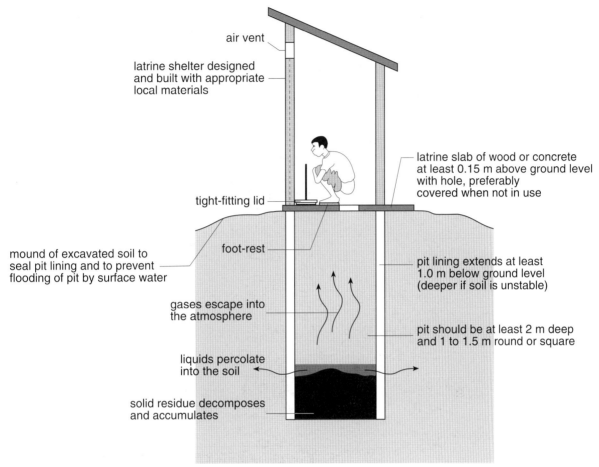

Figure 41   Diagram of a simple pit latrine [Rod Shaw, WEDC Publications]

The pit latrine should be sited downhill of settlements and water sources, and at least 50 m away from them. The superstructure of the pit latrine can be made with locally available materials, such as wood, mud and grass, thus keeping costs down.

12.4.2   The ventilated improved pit latrine (VIP)

This is an improved version of the simple pit latrine, designed to overcome the fly and odour problems of the latter. A vent pipe is incorporated into the design (Figure 42) to remove odorous gases from the pit. The pipe should be located outside the latrine, and should extend at least 50 cm above the latrine superstructure. It should be painted black to increase solar heating of the air in the pipe, causing it to rise by convection. The rise of the air through the pipe draws the smells away from the pit. Any flies which enter or breed in the pit are attracted by the relatively bright light at the end of the vent pipe. The open end of the pipe is covered with a gauze mesh or fly-proof netting, designed to prevent flies entering the pit and to trap any flies trying to leave. The flies eventually die and fall back into the pit. The mesh should be of PVC-coated fibreglass, as the vent pipe gases are aggressive.

The interior of the latrine should be kept dark to deter flies but there should be a gap above the door to allow air to enter. Air flow can be increased by positioning the latrine such that its door faces the prevailing wind.

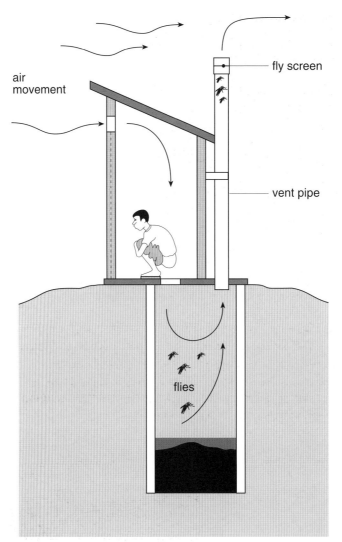

Figure 42   Diagram of a ventilated improved pit latrine [Rod Shaw, WEDC Publications]

While the VIP latrine has fewer fly and odour problems than a simple pit latrine, it is more difficult to construct, and is more expensive. The dark interior can deter young children from using it, and there is an increased odour outside.

12.4.3   Sullage

Stagnant pools of sullage, and also tap water which has spilt and run to waste, can be used as breeding grounds by mosquitoes and flies, and can be a danger to children. For minimal cost and minimal health risk, sullage should be disposed of as close as possible to its point of origin.

Natural drainage
The simplest method of disposal is to direct sullage to a flowing watercourse, downstream of any abstraction point. Sullage should never be discharged to stagnant ponds where it may become anaerobic and offensive.

Soakaways
Where soils are sufficiently permeable, soakaways (Figure 43) are appropriate. They are commonly 2–5 m deep and 1–2.5 m in diameter. Wastewater soaks away through the base and sides of the soakaway. The soakaway is filled with large stones, blocks, etc. in order to support the walls. The size of the soakaway should be fixed after the infiltration capacity of the soil has been determined. Seasonal variations in infiltration rate should be considered.

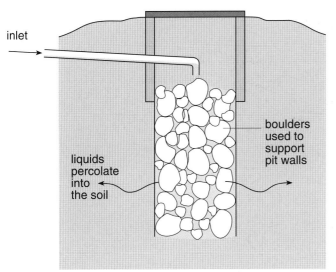

Figure 43   Diagram of a covered soakaway [Rod Shaw, WEDC Publications]

Sullage with a high solids content should be strained (e.g. with a woven sacking strainer) to prevent the soil pores becoming clogged. Removal of grease with a grease trap (Figure 44) is also advisable. The grease trap illustrated shows a strainer to remove solids, and a series of baffles to trap any grease which floats to the surface. Clean water travels underneath the baffles, and out through the overflow. Grease traps should be emptied daily.

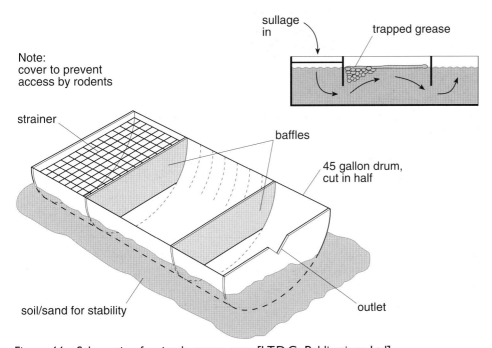

Figure 44   Schematic of a simple grease trap [I.T.D.G. Publications Ltd]

Where the ground is not suitable for a soakaway (e.g. in rocky or clay-rich areas) three options for dealing with sullage are:

- evaporation pans
- evapotranspiration beds
- irrigation.

Evaporation pans

Evaporation pans are shallow ponds (Figure 45) which hold water and allow it to evaporate. Such ponds are ideal for arid climates. They require a lot of land. Even with a high evaporation rate of 5 mm d^{-1}, 200 m^2 of surface area is required per cubic metre of liquid per day. The pans should be sited away from habitation to reduce the hazards due to mosquitoes and flies. They should only be used in areas where the mean evaporation rate is at least 4 mm d^{-1}, where rainfall is negligible, and where there is no viable alternative.

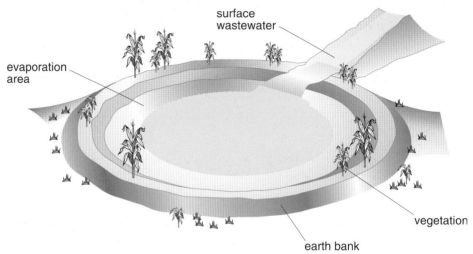

Figure 45   Diagram of an evaporation pan [Rod Shaw, WEDC Publications]

SAQ 64

(a) Which factors does evaporation rate depend on?

(b) If 200 m^2 is required per cubic metre of liquid, what is the depth of liquid when first put in?

Evapotranspiration beds

Evapotranspiration beds (Figure 46) rely on capillary action to draw water to the surface of shallow sand beds for evaporation to the atmosphere. Evapotranspiration beds are planted with grass or other vegetation which increases the movement of water from the root zone for transpiration through the leaves. The wastewater enters the bed through a system of distribution pipes surrounded by gravel (20–50 mm in diameter). A permeable filter cloth is placed over the gravel, and the bed is filled with sand and covered with a layer of topsoil in which grass is planted. The beds are as shallow as possible (not more than 1 m deep), to keep them aerobic and to prevent clogging. The performance of evapotranspiration beds depends on several factors (e.g. climate, wastewater flow, type of soil, vegetation). Hydraulic loading rates of up to 10 litres m^{-2} d^{-1} have been applied.

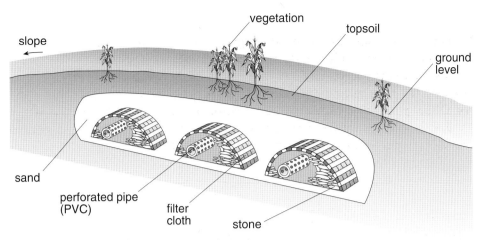

Figure 46 Diagram of an evapotranspiration bed [Rod Shaw, WEDC Publications]

Irrigation

Irrigation with sullage is a possibility but the health risks must be assessed, especially if vegetables are to be grown. Fast-growing trees such as banana or papaya can be planted in the drainage channels.

Rainwater run-off

Rainwater run-off in a refugee camp has to be controlled because it can cause erosion, especially if the flow is high. Any run-off is also likely to contain pollutants. Drainage channels should be constructed to lead the rainwater away from the campsite.

APPENDIX 1: CHARACTERISTICS OF THE QUALITY OF SURFACE WATER INTENDED FOR THE ABSTRACTION OF DRINKING WATER

Table A1

	Parameters		A1 G	A1 I	A2 G	A2 I	A3 G	A3 I
1	pH		6.5 to 8.5		5.5 to 9		5.5 to 9	
2	coloration (after simple filtration)	mg l^{-1} Pt/Co scale	10	20(O)	50	100(O)	50	200(O)
3	total suspended solids	mg l^{-1} SS	25					
4	temperature	°C	22	25(O)	22	25(O)	22	25(O)
5	conductivity	mS cm^{-1} at 20 °C	1000		1000		1000	
6	odour	(dilution factor at 25 °C)	3		10		20	
7[a]	nitrates	mg l^{-1} NO$_3$	25	50(O)		50(O)		50(O)
8[b]	fluorides	mg l^{-1} F	0.7 to 1	1.5	0.7 to 1.7		0.7 to 1.7	
9	total extractable organic chlorine	mg l^{-1} Cl						
10[a]	dissolved iron	mg l^{-1} Fe	0.1	0.3	1	2	1	
11[a]	manganese	mg l^{-1} Mn	0.05		0.1		1	
12	copper	mg l^{-1} Cu	0.02		0.05(O)	0.05		1
13	zinc	mg l^{-1} Zn	0.5	3	1	5	1	5
14	boron	mg l^{-1} B	1		1		1	
15	beryllium	mg l^{-1} Be						
16	cobalt	mg l^{-1} Co						
17	nickel	mg l^{-1} Ni						
18	vanadium	mg l^{-1} V						
19	arsenic	mg l^{-1} As	0.01	0.05		0.05	0.05	0.1
20	cadmium	mg l^{-1} Cd	0.001	0.005	0.001	0.005	0.001	0.005
21	total chromium	mg l^{-1} Cr		0.05		0.05		0.05
22	lead	mg l^{-1} Pb		0.05		0.05		0.05
23	selenium	mg l^{-1} Se		0.01		0.01		0.01
24	mercury	mg l^{-1} Hg	0.0005	0.001	0.0005	0.001	0.0005	0.001
25	barium	mg l^{-1} Ba		0.1		1		1
26	cyanide	mg l^{-1} CN		0.05		0.05		0.05
27	sulphates	mg l^{-1} SO$_4$	150	250	150	250(O)	150	250(O)
28	chlorides	mg l^{-1} Cl	200		200		200	
29	surfactants (reacting with methylene blue)	mg l^{-1} (laurylsulphate)	0.2		0.2		0.5	
30[ac]	phosphates	mg l^{-1} P$_2$O$_5$	0.4		0.7		0.7	

	Parameters		A1 G	A1 I	A2 G	A2 I	A3 G	A3 I
31	phenols (phenol index) paranitraniline 4-aminoantipyrine	mg l^{-1} C$_6$H$_5$OH		0.001	0.001	0.005	0.01	0.1
32	dissolved or emulsified hydrocarbons (after extraction by petroleum ether)	mg l^{-1}			0.05	0.2	0.5	1
33	polycyclic aromatic hydrocarbons	mg l^{-1}		0.0002		0.0002		0.001
34	total pesticides (parathion, BHC, dieldrin)	mg l^{-1}		0.001		0.0025		0.005
35^a	chemical oxygen demand (COD)	mg l^{-1} O$_2$					30	
36^a	dissolved oxygen saturation rate	% O$_2$	$\geqslant 70$		$\geqslant 50$		$\geqslant 30$	
37^a	biochemical oxygen demand (BOD$_5$) (at 20 °C without nitrification)	mg l^{-1} O$_2$	$\leqslant 3$		$\leqslant 5$		$\leqslant 7$	
38	nitrogen by Kjeldahl method (except NO$_3$)	mg l^{-1} N	1		2		3	
39	ammonia	mg l^{-1} NH$_4$	0.05		1	1.5	2	4(O)
40	substances extractable with chloroform (SEC)	mg l^{-1} SEC	0.1		0.2		0.5	
41	total organic carbon	mg l^{-1} C						
42	residual organic carbon after flocculation and membrane filtration (5 μm) TOC	mg l^{-1} C						
43	total coliforms 37 °C	/100 ml	50		5000		50 000	
44	faecal coliforms	/100 ml	20		2000		20 000	
45	faecal streptococci	/100 ml	20		1000		10 000	
46	*Salmonella*		not present in 5000 ml		not present in 1000 ml			

Source: EU Directive 75/440/EEC.

I, Mandatory; G, Guide; (O) The directive may be waived in exceptional meteorological or geographical conditions.

^{a}The directive may be waived in the case of surface water in shallow lakes, or virtually stagnant surface water, providing the depth of the lake does not exceed 20 m, the lake has an exchange of water slower than 1 year, and without a discharge of wastewater into the water body.

^{b}The values given are upper limits set in relation to the mean annual temperature (high and low).

^{c} This parameter has been included to satisfy the ecological requirements of certain types of environment.

Definition of the standard methods of treatment for transforming surface water of categories A1, A2 and A3 into drinking water:

Category A1: Simple physical treatment and disinfection, e.g. rapid filtration and disinfection.

Category A2: Normal physical treatment, chemical treatment and disinfection, e.g. prechlorination, coagulation, flocculation, decantation, filtration, disinfection (final chlorination).

Category A3: Intensive physical and chemical treatment, extended treatment and disinfection, e.g. chlorination to break-point, coagulation, flocculation, decantation, filtration, adsorption (activated carbon), disinfection (ozone, final chlorination).

APPENDIX 2: EU DIRECTIVE ON THE QUALITY OF WATER INTENDED FOR HUMAN CONSUMPTION (98/83/EC)

This directive covers all water for domestic use and water used by the food industry where this affects the final product (and thus consumers' health). It does not apply to water used for agricultural purposes, natural mineral water or medicinal waters. The parameters in the directive will be reviewed at least every five years.

The new directive splits the quality parameters into two categories: Mandatory microbiological and chemical parameters, and Indicator parameters.

Mandatory parameters

For tap water, there are two microbiological and 26 chemical parameters (Tables A2 and A3). Instead of 'guide levels' and 'maximum admissible concentration' (as in the old directive), there are now 'parametric values' (PVs).

Table A2   Microbiological parameters

Parameter	Parametric value (number per 100 ml)
Escherichia coli (E. coli)	0
Enterococci	0

The following applies to water offered for sale in bottles or containers:

Parameter	Parametric value
Escherichia coli (E. coli)	0 per 250 ml
Enterococci	0 per 250 ml
Pseudomonas aeruginosa	0 per 250 ml
Colony count at 22 °C	10 per ml
Colony count at 37 °C	20 per ml

Table A3   Chemical parameters

Parameter	Parametric value	Unit	Notes
Acrylamide	0.10	$\mu g \, l^{-1}$	Note 1
Antimony	5.0	$\mu g \, l^{-1}$	
Arsenic	10	$\mu g \, l^{-1}$	
Benzene	1.0	$\mu g \, l^{-1}$	
Benzo(a)pyrene	0.010	$\mu g \, l^{-1}$	
Boron	1.0	$mg \, l^{-1}$	
Bromate	10	$\mu g \, l^{-1}$	Note 2
Cadmium	5.0	$\mu g \, l^{-1}$	
Chromium	50	$\mu g \, l^{-1}$	
Copper	2.0	$mg \, l^{-1}$	Note 3
Cyanide	50	$\mu g \, l^{-1}$	

1,2-dichloroethane	3.0	μg l^{-1}	
Epichlorohydrin	0.10	μg l^{-1}	Note 1
Fluoride	1.5	mg l^{-1}	
Lead	10	μg l^{-1}	Notes 3 and 4
Mercury	1.0	μg l^{-1}	
Nickel	20	μg l^{-1}	Note 3
Nitrate	50	mg l^{-1}	Note 5
Nitrite	0.50	mg l^{-1}	Note 5
Pesticides	0.10	μg l^{-1}	Notes 6 and 7
Pesticides – total	0.50	μg l^{-1}	Notes 6 and 8
Polycyclic aromatic hydrocarbons	0.10	μg l^{-1}	Sum of concentrations of specified compounds; Note 9
Selenium	10	μg l^{-1}	
Tetrachloroethene and trichloroethene	10	μg l^{-1}	Sum of concentrations of specified parameters
Trihalomethanes – total	100	μg l^{-1}	Sum of concentrations of specified compounds; Note 10
Vinyl chloride	0.50	μg l^{-1}	Note 1

Note 1:   The parametric value refers to the residual monomer concentration in the water as calculated according to specifications of the maximum release from the corresponding polymer in contact with the water.

Note 2:   Where possible, without compromising disinfection, Member States should strive for a lower value.
For the water referred to in Article 6(1)(a), (b) and (d) [see below], the value must be met, at the latest, 10 calendar years after the entry into force of the directive. The parametric value for bromate from five years after the entry into force of the directive until 10 years after its entry into force is 25 μg l^{-1}.

Article 6
Point of compliance
The parametric values set in accordance with Article 5 shall be complied with:
(a)   in the case of water supplied from a distribution network, at the point, within premises or an establishment, at which it emerges from the taps that are normally used for human consumption;
(b)   in the case of water supplied from a tanker, at the point at which it emerges from the tanker;
(c)   in the case of water put into bottles or containers intended for sale, at the point at which the water is put into bottles or containers;
(d)   in the case of water used in food production undertaking, at the point where the water is used in the undertaking.

Note 3:   The value applies to a sample of water intended for human consumption obtained by an adequate sampling method at the tap and taken so as to be representative of a weekly average value ingested by consumers. Where appropriate, the sampling and monitoring methods must be applied in a harmonised fashion to be drawn up in accordance with Article 7(4). Member States must take account of the occurrence of peak levels that may cause adverse effects on human health.

Note 4:   For water referred to in Article 6(1)(a), (b) and (d), the value must be met, at the latest, 15 calendar years after the entry into force of this directive. The parametric value for lead from five years after the entry into force of this directive, until 15 years after its entry into force is 25 μg l^{-1}.

Member States must ensure that all appropriate measures are taken to reduce the concentration of lead in water intended for human consumption as much as possible during the period needed to achieve compliance with the parametric value.

When implementing the measures to achieve compliance with that value, Member States must progressively give priority where lead concentrations in water intended for human consumption are highest.

Note 5:   Member States must ensure that the condition that [nitrate]/50 + [nitrate]/3 $\leqslant$ 1, the square brackets signifying the concentrations in mg l^{-1} for nitrate (NO$_3$) and nitrite (NO$_2$), is complied with and that the value of 0.10 mg l^{-1} for nitrites is complied with ex-water treatment works.

Note 6:   'Pesticides' means:
    organic insecticides,
    organic herbicides,
    organic fungicides,
    organic nematocides,
    organic acaricides,
    organic algicides,
    organic rodenticides,
    organic slimicides,
    related products, (*inter alia*, growth regulators)
and their relevant metabolites, degradation and reaction products.
Only those pesticides which are likely to be present in a given supply need be monitored.

Note 7:   The parametric value applies to each individual pesticide. In the case of aldrin, dieldrin, heptachlor and heptachlor epoxide the parametric value is 0.030 μg l^{-1}.

Note 8:   'Pesticides–total' means the sum of all individual pesticides detected and quantified in the monitoring procedure.

Note 9:   The specified compounds are:
    benzo(b)fluoranthene,
    benzo(k)fluoranthene,
    benzo(ghi)perylene,
    indeno(1,2,3-cd)pyrene

Note 10:   Where possible, without compromising disinfection, Member States should strive for a lower value.

The specified compounds are: chloroform, bromoform, dibromochloromethane, bromodichloromethane.

For the water referred to in Article 6(1)(a), (b) and (d), the value must be met, at the latest, 10 calendar years after the entry into force of this directive. The parametric value for total THMs from five years after the entry into force of this directive until 10 years after its entry into force is 150 μg l^{-1}.

Member States must ensure that all appropriate measures are taken to reduce the concentration of THMs in water intended for human consumption as much as possible during the period needed to achieve compliance with the parametric value.

When implementing the measures to achieve this value, Member States must progressively give priority to those areas where THM concentrations in water intended for human consumption are highest.

Indicator parameters

These parameters were set for monitoring purposes.

This is a new concept. Generally speaking, the PVs for these parameters are not based on health considerations and are non-binding. When the PVs are exceeded, remedial action is only necessary if a Member State judges there to be risk to health. Most of the Indicator parameters are included in the list of parameters to be analysed in Check monitoring. This has the aim of providing information on the organoleptic and microbiological quality of supplied water, as well as on the effectiveness of treatment. The Indicator parameters are listed in Table A3 below. A fair number of the existing EC standards have been turned into Indicator parameters in the new directive, or omitted altogether. The Department for Environment, Food and Rural Affairs, however, proposes to retain most of the existing EC standards as binding limits. This means, for instance, that standards for aluminium, iron, manganese, colour and turbidity intended to prevent the supply of discoloured water, will be retained.

Table A3   Indicator parameters

Parameter	Parametric value	Unit	Notes
Aluminium	200	μg l^{-1}	
Ammonium	0.50	mg l^{-1}	
Chloride	250	mg l^{-1}	Note 1
Clostridium perfringens (including spores)	0	number per 100 ml	Note 2
Colour	Acceptable to consumers and no abnormal change		
Conductivity	2500	μS cm^{-1} at 20 °C	Note 1
Hydrogen ion concentration	> 6.5 and > 9.5	pH units	Notes 1 and 3
Iron	200	μg l^{-1}	
Manganese	50	μg l^{-1}	
Odour	Acceptable to consumers and no abnormal change		
Oxidisability	5.0	mg l^{-1} O$_2$	Note 4
Sulphate	250	mg l^{-1}	Note 1
Sodium	200	mg l^{-1}	
Taste	Acceptable to consumers and no abnormal change		
Colony count at 22 °C	No abnormal change		
Coliform bacteria	0	number per 100 ml	Note 5
Total organic carbon (TOC)	No abnormal change		Note 6
Turbidity	Acceptable to consumers and no abnormal change		Note 7
Radioactivity			
Tritium	100	Bq l^{-1}	Notes 8 and 10
Total indicative dose	0.10	mSv year^{-1}	Notes 9 and 10

Note 1:   The water should not be aggressive.

Note 2:   This parameter need not be measured unless the water originates from or is influenced by surface water. In the event of non-compliance with this parametric value, the Member State concerned must investigate the supply to ensure that there is no potential danger to human health arising from the presence of pathogenic micro-organisms, e.g. *Cryptosporidium*. Member States must include the results of all such investigations in the reports they must submit under Article 13(2).

Note 3:   For still water put into bottles or containers, the minimum value may be reduced to 4.5 pH units.
For water put into bottles or containers which is naturally rich in or artificially enriched with carbon dioxide, the minimum value may be lower.

Note 4:   This parameter need not be measured if the parameter TOC is analysed.

Note 5:   For water put into bottles or containers the unit is number per 250 ml.

Note 6:   This parameter need not be measured for supplies of less than 10 000 m^3 a day.

Note 7:   In the case of surface water treatment, Member States should strive for a parametric value not exceeding 1.0 NTU (nephelometric turbidity unit) in the water ex-treatment works.

Note 8:   Monitoring frequencies to be set later in Annex II.

Note 9:   Excluding tritium, potassium-40, radon and radon decay products; monitoring frequencies, monitoring methods and the most relevant locations for monitoring points to be set later in Annex II.

Note 10:   1   The proposals required by Note 8 on monitoring frequencies, and Note 9 on monitoring frequencies, monitoring methods and the most relevant locations for monitoring points in Annex II shall be adopted in accordance with the procedure laid down in Article 12. When elaborating these proposals the Commission shall take into account *inter alia* the relevant provisions under existing legislation or appropriate monitoring programmes including monitoring results as derived from them. The Commission shall submit these proposals at the latest within 18 months following the date referred to in Article 18 of the directive.

2   A Member State is not required to monitor drinking water for tritium or radioactivity to establish total indicative dose where it is satisfied that, on the basis of other monitoring carried out, the levels of tritium of the calculated total indicated dose are well below the parametric value. In that case, it shall communicate the grounds for its decision to the Commission, including the results of this other monitoring carried out.

APPENDIX 3: THE WATER SUPPLY (WATER QUALITY) REGULATIONS 2000

Prescribed concentrations and values

Table A4   Microbiological parameters

Part I: Directive requirements

Item	Parameters	Concentration or value (maximum)	Units of measurement	Point of compliance
1.	Enterococci	0	number per 100 ml	Consumers' taps
2.	*Escherichia coli* (*E. coli*)	0	number per 100 ml	Consumers' taps

Part II: National requirements

Item	Parameters	Concentration or value (maximum)	Units of measurement	Point of compliance
1.	Coliform bacteria	0	number per 100 ml	Service reservoirs* and water treatment works
2.	*Escherichia coli* (*E. coli*)	0	number per 100 ml	Service reservoirs and water treatment works

Note: *Compliance required as to 95% of samples from each service reservoir (regulation 4(6)).

Table A5   Chemical parameters

Part I: Directive requirements

Item	Parameters	Concentration or value (maximum)	Units of measurement	Point of compliance
1.	Acrylamide	0.10	$\mu g\ l^{-1}$	**(i)**
2.	Antimony	5.0	$\mu g\ Sb\ l^{-1}$	Consumers' taps
3.	Arsenic	10	$\mu g\ As\ l^{-1}$	Consumers' taps
4.	Benzene	1.0	$\mu g\ l^{-1}$	Consumers' taps
5.	Benzo(a)pyrene	0.010	$\mu g\ l^{-1}$	Consumers' taps
6.	Boron	1.0	$mg\ B\ l^{-1}$	Consumers' taps
7.	Bromate	10	$\mu g\ BrO_3\ l^{-1}$	Consumers' taps
8.	Cadmium	5.0	$\mu g\ Cd\ l^{-1}$	Consumers' taps
9.	Chromium	50	$\mu g\ Cr\ l^{-1}$	Consumers' taps
10.	Copper **(ii)**	2.0	$mg\ Cu\ l^{-1}$	Consumers' taps
11.	Cyanide	50	$\mu g\ CN\ l^{-1}$	Consumers' taps
12.	1,2dichloroethane	3.0	$\mu g\ l^{-1}$	Consumers' taps
13.	Epichlorohydrin	0.10	$\mu g\ l^{-1}$	**(i)**
14.	Fluoride	1.5	$mg\ F\ l^{-1}$	Consumers' taps
15.	Lead **(ii)**	(a) 25, from 25th December 2003 until immediately before 25th December 2013	$\mu g\ Pb\ l^{-1}$	Consumers' taps
		(b) 10, on and after 25th December 2013	$\mu g\ Pb\ l^{-1}$	Consumers' taps
16.	Mercury	1.0	$\mu g\ Hg\ l^{-1}$	Consumers' taps
17.	Nickel **(ii)**	20	$\mu g\ Ni\ l^{-1}$	Consumers' taps
18.	Nitrate **(iii)**	50	$mg\ NO_3\ l^{-1}$	Consumers' taps
19.	Nitrite **(iii)**	0.50	$mg\ NO_2\ l^{-1}$	Consumers' taps
		0.10		Treatment works
20.	Pesticides **(iv)(v)**			
	Aldrin	0.030	$\mu g\ l^{-1}$	Consumers' taps
	Dieldrin	0.030	$\mu g\ l^{-1}$	Consumers' taps
	Heptachlor	0.030	$\mu g\ l^{-1}$	Consumers' taps
	Heptachlor epoxide	0.030	$\mu g\ l^{-1}$	Consumers' taps
	Other pesticides	0.10	$\mu g\ l^{-1}$	Consumers' taps
21.	Pesticides: total **(vi)**	0.50	$\mu g\ l^{-1}$	Consumers' taps
22.	Polycyclic aromatic hydrocarbons **(vii)**	0.10	$\mu g\ l^{-1}$	Consumers' taps
23.	Selenium	10	$\mu g\ Se\ l^{-1}$	Consumers' taps
24.	Tetrachloroethene and trichloroethene **(viii)**	10	$\mu g\ l^{-1}$	Consumers' taps
25.	Trihalomethanes: total **(ix)**	100	$\mu g\ l^{-1}$	Consumers' taps
26.	Vinyl chloride	0.50	$\mu g\ l^{-1}$	**(i)**

Notes:

(i)   The parametric value refers to the residual monomer concentration in the water as calculated according to specifications of the maximum release from the corresponding polymer in contact with the water. This is controlled by product specification.

(ii)   See also regulation 6(6).

(iii)   See also regulation 4(2)(d).

(iv)   See the definition of 'pesticides and related products' in regulation 2.

(v)   The parametric value applies to each individual pesticide.

(vi)   'Pesticides: total' means the sum of the concentrations of the individual pesticides detected and quantified in the monitoring procedure.

(vii)   The specified compounds are:

benzo(b)fluoranthene

benzo(k)fluoranthene

benzo(ghi)perylene

indeno(1,2,3-cd)pyrene.

The parametric value applies to the sum of the concentrations of the individual compounds detected and quantified in the monitoring process.

(viii)   The parametric value applies to the sum of the concentrations of the individual compounds detected and quantified in the monitoring process.

(ix)   The specified compounds are:

chloroform

bromoform

dibromochloromethane

bromodichloromethane.

The parametric value applies to the sum of the concentrations of the individual compounds detected and quantified in the monitoring process.

Part II: National requirements

Item	Parameters	Concentration or value (maximum unless otherwise stated)	Units of measurement	Point of compliance
1.	Aluminium	200	μg Al l^{-1}	Consumers' taps
2.	Colour	20	mg l^{-1} Pt/Co	Consumers' taps
3.	Hydrogen ion	10.0	pH value	Consumers' taps
		6.5 (minimum)	pH value	
4.	Iron	200	μg Fe l^{-1}	Consumers' taps
5.	Manganese	50	μg Mn l^{-1}	Consumers' taps
6.	Odour	3 at 25 °C	Dilution number	Consumers' taps
7.	Sodium	200	mg Na l^{-1}	Consumers' taps
8.	Taste	3 at 25 °C	Dilution number	Consumers' taps
9.	Tetrachloromethane	3	μg l^{-1}	Consumers' taps
10.	Turbidity	4	NTU	Consumers' taps

Indicator parameters

Item	Parameters	Specification concentration or value (maximum or state)	Units of measurement	Point of monitoring
1.	Ammonium	0.50	mg NH_4 l^{-1}	Consumers' taps
2.	Chloride **(i)**	250	mg Cl l^{-1}	Supply point*
3.	*Clostridium perfringens* (including spores)	0	Number per 100 ml	Supply point*
4.	Coliform bacteria	0	Number per 100 ml	Consumers' taps
5a.	Colony counts	No abnormal change	Number per 1 ml at 22 °C	Consumers' taps, service reservoirs and treatment works
5b.	Colony counts	No abnormal change	Number per 1 ml at 37 °C	Consumers' taps, service reservoirs and treatment works
6.	Conductivity **(i)**	2500	μS cm^{-1} at 20 °C	Supply point*
7.	Sulphate **(i)**	250	mg SO_4 l^{-1}	Supply point *
8.	Total indicative dose (for radioactivity) **(ii)**	0.10	mSv $year^{-1}$	Supply point*
9.	Total organic carbon (TOC)	No abnormal change	mg C l^{-1}	Supply point*
10.	Tritium for radioactivity	100	Bq l^{-1}	Supply point*
11.	Turbidity	1	NTU	Treatment works

* May be monitored from samples of water leaving treatment works or other supply point, as no significant change during distribution.

Notes:

(i) The water should not be aggressive.

(ii) Excluding tritium, potassium-40, radon and radon decay products.

APPENDIX 4: MONITORING

Source: From EU Directive 75/440/EEC.

Table A6   Parameters to be analysed

1   Check monitoring

The purpose of check monitoring is regularly to provide information on the organoleptic and microbiological quality of the water supplied for human consumption as well as information on the effectiveness of drinking-water treatment (particularly of disinfection) where it is used, in order to determine whether or not water intended for human consumption complies with the relevant parametric values laid down in this directive.

The following parameters must be subject to check monitoring. Member States may add other parameters to this list if they deem it appropriate.

Aluminium (Note 1)

Ammonium

Colour

Conductivity

Clostridium perfringens (including spores) (Note 2)

Escherichia coli (*E. coli*)

Hydrogen ion concentration

Iron (Note 1)

Nitrite (Note 3)

Odour

Pseudomonas aeruginosa (Note 4)

Taste

Colony count 22 °C and 37 °C (Note 4)

Coliform bacteria

Turbidity

Note 1: Necessary only when used as flocculant (*)

Note 2: Necessary only if the water originates from or is influenced by surface water (*)

Note 3: Necessary only when chloramination is used as a disinfectant (*)

Note 4: Necessary only in the case of water offered for sale in bottles or containers.

(*) In all other cases, the parameters are in the list for audit monitoring.

2   Audit monitoring

The purpose of audit monitoring is to provide the information necessary to determine whether or not all of the directive's parametric values are being complied with. All parameters set in accordance with Articles 5(2) and (3) must be subject to audit monitoring unless it can be established by the competent authorities, for a period of time to be determined by them, that a parameter is not likely to be present in a given supply in concentrations which could lead to the risk of a breach of the relevant parametric value. This paragraph does not apply to the parameters for radioactivity, which, subject to Notes 8, 9 and 10 in Annex I, Part C, will be monitored in accordance with monitoring requirements adopted under Article 12.

Table B   Minimum frequency of sampling and analysis for water intended for human consumption supplied from a distribution network or from a tanker or used in a food-production undertaking

Member States must take samples at the points of compliance as defined in Article 6(1) to ensure that water intended for human consumption meets the requirements of the directive. However, in the case of a distribution network, a Member State may take samples within the supply zone or at the treatment works for particular parameters if it can be demonstrated that there would be no adverse change to the measured value of the parameters concerned.

Volume of water distributed or produced each day within a supply zone (m^3) (Notes 1 and 2)		Check monitoring number of samples per year (Notes 3, 4 and 5)	Audit monitoring number of samples per year (Notes 3 and 5)
	⩽ 100	(Note 6)	(Note 6)
> 100	⩽ 1000	4	1
> 1000	⩽ 10 000	4 + 3 for each 1000 m^3 per day and part thereof of the total volume	1 + 1 for each 3300 m^3 per day and part thereof of the total volume
> 10 000	⩽ 100 000		3 + 1 for each 10 000 m^3 per day and part thereof of the total volume
> 100 000			10 + 1 for each 25 000 m^3 per day and part thereof of the total volume

Note 1: A supply zone is a geographically defined area within which water intended for human consumption comes from one or more sources and within which water quality may be considered as being approximately uniform.

Note 2: The volumes are calculated as averages taken over a calendar year. A Member State may use the number of inhabitants in a supply zone instead of the volume of water to determine the minimum frequency, assuming a water consumption of 200 litres per day per capita.

Note 3: In the event of intermittent short-term supply the monitoring frequency of water distributed by tankers is to be decided by the Member State concerned.

Note 4: For the different parameters in Annex I, a Member State may reduce the number of samples specified in the table if:

    (a) the values of the results obtained from samples taken during a period of at least two successive years are constant and significantly better than the limits laid down in Annex I, and

    (b) no factor is likely to cause a deterioration of the quality of the water.

       The lowest frequency applied must not be less than 50% of the number of samples specified in the table except in the particular case of Note 6.

Note 5: As far as possible, the number of samples should be distributed equally in time and location.

Note 6: The frequency is to be decided by the Member State concerned.

APPENDIX 5: TRADE EFFLUENT DISCHARGED TO THE SEWER: RECOMMENDED GUIDELINES FOR CONTROL AND CHARGING

Source: CBI/Regional Water Authorities (1986), published by Water Services Association

Objectives of trade effluent control

1 To prevent trade effluent discharge to sewers causing:

(a) damage or harm to the sewerage system and the personnel employed therein.

(b) interference with the effective and economic treatment of the mixed sewage by the processes employed at the sewage works.

(c) the products of that treatment, in the form of effluents or residues, to have unacceptable effects on water resources or the environment generally.

(d) unacceptable storm sewage discharges to watercourses.

The achievement of these first objectives might require segregation of noxious process streams or pretreatment of the effluent to be undertaken by the trader, before discharge to the sewer, to reduce the concentrations of certain substances.

2 To provide data regarding the volume, rate of discharge, nature and composition of trade effluent discharges to sewers for use in the design of future sewerage and sewage treatment works, and in ensuring the proper management of the quality of water resources.

3 To ensure that the trader pays a reasonable charge having regard to the costs incurred by the Water Service Companies for providing that part of the service which is used for the reception, conveyance, treatment and disposal of trade effluent without showing undue discrimination or undue preference.

Operation of scheme of control

4 All trade effluents discharged to sewer should be controlled in accordance with these Guidelines, with the following exceptions:

(a) Discharges under existing Agreements until such times as these may be superseded.

(b) Discharges under existing Consents. The recommended procedure should be adopted in these cases as far as possible until such time as a new Consent or Direction can be issued.

(c) Premises which have the benefit of Crown privilege should be requested to comply with the normal Consent conditions as if they were subject to the general trade effluent control legislation. (Please note that under the Environment Act 1995, Crown Immunity was removed from 1 April 1996.)

(d) Discharge of radioactive substances controlled by the provisions of the Radioactive Substances Act 1960. A system of notification of emergency discharges to sewers will be necessary to prevent hazards to personnel employed in sewers.

Consent conditions

5   Such limitations as are imposed by the Water Service Companies should be to ensure that no hazard occurs in the sewer or neighbouring premises, that no damage can be caused to the sewer, that the waste can be treated properly at the sewage works without deleterious effects and that any products of treatment (e.g. effluents, sludges) should not have unacceptable effects on water resources or the environment generally. The Water Service Company must be aware of its own responsibilities under Common Law and, depending on its sludge disposal practices, to agriculture and under international conventions such as those covered by the Dumping at Sea Act 1974. Particular regard must be had to the Health and Safety at Work etc. Act 1974.

6   All practical means of reducing at source the volume and polluting characteristics of trade effluents should be encouraged, but the Consent Conditions applied should be only to the degree necessary to ensure compliance with the principles outlined in the preceding paragraph, or where there is no early possibility of providing adequate sewerage or treatment capacity.

7   This does not preclude the trader from giving further pretreatment than actually required to meet the conditions imposed by the Water Service Company, e.g. in those circumstances where he considers it economical to reduce the Water Service Company's charges for subsequent treatment.

Directions

8   Directions are issued to vary the conditions laid down in a Consent issued under the provisions of Section 2(3) of the Public Health (Drainage of Trade Premises) Act 1937, or a Consent under Section 43(3) of the Control of Pollution Act 1974.

9(a) Directions varying the conditions laid down in a Consent or previous Direction may not normally be issued within two years from the date of the Consent or previous Direction without the trader's written consent. When Section 45 of the 1974 Act is in force, for the protection of persons likely to be affected by the discharge, the Water Service Company may consider it urgently necessary to vary conditions, in which case compensation might then become payable by the Water Service Company.

9(b) Each Consent should, therefore, be regularly reviewed in sufficient time to enable any appropriate Direction to be issued every two years.

10   Charges should recover the costs incurred in treatment and disposal of trade effluent on the basis of estimates of relevant expenditure. Costings used in the formula shall be regionally based.

Charging for trade effluents

11   Positive measurement of the volume of trade effluent discharge is generally required. In cases where this is not practicable, metered process volumes and appropriate allowances for domestic use and process losses should be used when such flows are included in the total metered water volume intake.

12   The frequency of sampling for charging purposes should be based on a reasonable compromise between the cost of sampling and analysis in relation to total discharge costs, recognising the size and variability of the discharge.

13   The oxidation charge should be based on COD (chemical oxygen demand from acidified dichromate). In some cases the use of an alternative oxidation factor may be applied following joint investigation and agreement.

14   The biological oxidation process produces, from soluble constituents of the sewage, secondary sludges in quantities related to the oxidation load on the plant and the cost of treatment and disposal of these sludges should be allocated to the oxidation load. It will generally be necessary to apportion the total cost of sludge treatment in relation to the proportions of primary and secondary sludge solids produced during the whole sewage treatment process and if separate costings for secondary sludge treatment are not available, one-third of the sludge costs should be allocated to the biological charge.

15   The biological oxidation process also includes items of plant, such as final settlement tanks, whose costs are volume-related. These costs should be recovered on a volume basis.

16   There should be no charge for toxic constituents as their concentrations should be so limited by the Consent Conditions as to render the waste capable of satisfactory treatment and disposal by the Water Service Company.

17   Trade effluents should be charged on the following type of formula based on samples taken over the normal working week:

$$C = R + V + \frac{O_t}{O_s}B + \frac{S_t}{S_s}S$$

where

C is the total charge per cubic metre of trade effluent

R is the reception and conveyance charge per cubic metre of sewage

V is the volumetric and primary treatment cost per cubic metre of sewage

O_t is the COD (in mg 1^{-1}) of the trade effluent after one hour quiescent settlement

O_s is the COD (in mg 1^{-1}) of settled sewage (standard strength)

B is the biological oxidation cost per cubic metre of settled sewage

S_t is the total suspended solids (in mg 1^{-1}) of the trade effluent

S_s is the total suspended solids (in mg 1^{-1}) of crude sewage (standard strength)

S is the treatment and disposal costs of primary sludge per cubic metre of sewage.

The settlement period should be one hour. In certain cases where the discharge is significantly different to pH 7, an adjustment may be appropriate.

[The standard strength factors (O_s and S_s) differ between the various Water Service Companies.]

In cases where the measurement of total suspended solids provides an unsuitable means for assessing the solids removed by settlement, an alternative method may be applied following joint investigation and agreement.

In cases where the present practice is to use settleable solids, this may continue.

18   It is recommended that, in order to assess the total flow and pollutional load of mixed sewage received at the Water Service Company's treatment works, an average based on the best available data should be used to avoid unreasonable variations.

19   Such fixed costs of the sewerage system which are deemed to be incurred by traders should be recovered from the traders in a manner consistent with the Water Service Company's charging policy, which may be a standing charge or a volume charge. Variable costs of the sewerage system should be recovered from traders through the unit cost R (reception and conveyance).

20   The unit cost for the term V (volumetric and primary treatment) should be derived from the net annual revenue expenditure including financing charges on capital on:

   (a)   all pumping stations with rising mains discharging directly to sewage treatment works,

   (b)   all inlet works, including screening, comminution, grit removal and preaeration,

   (c)   all primary settlement units other than storm treatment works,

   (d)   tertiary treatment for reduction of the concentration of residual suspended solids,

   (e)   all outfalls for treated sewage.

21   The unit cost for term B (biological treatment) should be derived from the net annual revenue expenditure including financing charges on capital on:

   (a)   biological filtration plants and humus tanks, including recirculation, alternating double filtration and humus sludge pumping,

   (b)   activated sludge plants and final settling tanks, including return sludge pumping,

   (c)   the proportion of total sludge treatment and disposal costs associated with secondary sludge treatment and disposal.

   In accordance with paragraph 15, financing charges on capital costs related to secondary settlement tanks, included under (a) and (b), should be recovered on a volume-related basis. This may require a separate charging factor.

22   The unit cost for term S (primary sludge treatment and disposal only: see paragraph 14) should be derived from that portion of the total net annual revenue expenditure including financing charges on capital related to primary sludges on:

   (a)   pumping or otherwise conveying (e.g. by tanker) sludge to treatment and disposal,

   (b)   sludge dewatering and treatment, including digestion, conditioning, consolidation, drying, storage, incineration and disposal.

23   A separate charge for sea outfalls (including discharges to tidal waters) may be made for those trade effluents which are otherwise untreated beyond preliminary treatment and are disposed of through a sea outfall belonging to and maintained by the Water Service Company. It is recommended that charges should have regard to the costs incurred in providing and operating sea outfalls on the basis of relevant expenditure. Costings used in the formula shall be regionally based.

   The charge may include one or more of the following factors:

   R is the reception and conveyance charge per cubic metre (for short outfalls with no additional facilities).

V_m is the volumetric and preliminary treatment cost per cubic metre (for outfalls which include additional headwork, pumps, tanks, etc.).

M is the marine costs per cubic metre (for specified long sea outfalls and additional works, e.g. diffusers).

24 Otherwise unallocated charges, such as site charges, headquarters and other central costs should be allocated over the treatment stages in proportion to the expenditure under each heading.

25 With relatively small or weak discharges a scheme of 'minimum charges' may be instituted where the product of flow and unit charge (calculated from the above formula) is less than a figure to be decided by each Water Service Company from time to time.

26 Charges will be made only for those elements of the formula applicable to the reception and disposal of the discharge relating to the type of treatment actually given to the trade effluent.

27 In circumstances in which a capital contribution is obtained from the trader a reduction in the normal trade effluent charges would be appropriate related to the circumstances of the case.

28 It is recognised that the implementation of Section 52 of the Control of Pollution Act 1974 (applying to trade effluent discharges the provisions of Section 30 of the Water Act 1973) could have an impact on the future assessment of charges.

ANSWERS TO SELF-ASSESSMENT QUESTIONS

SAQ I

(a) Dissolved oxygen – the amount of oxygen dissolved in a stream, river or lake, usually expressed in mg l^{-1} or g m^{-3}, or as a percentage of saturation concentration.

(b) Oxygen deficit – the difference between the existing dissolved oxygen concentration and the saturation value.

(c) Biochemical oxygen demand – the amount of oxygen consumed when a water sample is incubated for 5 days in darkness at 20 °C. Expressed in mg l^{-1} or g m^{-3}. The oxygen is used up in the biodegradation of organics by micro-organisms.

SAQ 2

$$ASPT = \frac{96}{6} = 16$$

$$EQI_{ASPT} = \frac{16}{20} = 0.8$$

$$EQI_{No. \, of \, taxa} = \frac{6}{8} = 0.75$$

From Table 3, the EQI_{ASPT} value indicates the river is Grade C and for the $EQI_{no. \, of \, taxa}$ it is Grade B. The lower of the two grades has to be applied. Hence the river is Grade C.

SAQ 3

(a) RIVPACS – River Invertebrate Prediction and Classification System. A computer model that predicts the type of invertebrate community that can be expected in a river, based on the existing natural features and assuming the river is not affected by pollution.

(b) ASPT – Average Score Per Taxon. This is the BMWP score for a given sampling site, divided by the number of taxa present. An ASPT greater than 4 generally indicates good water quality.

(c) EQI – Ecological Quality Index. The ratio of the observed community status to that predicted for a site. For example, the EQI for the BMWP score would be

$$EQI_{BMWP} = \frac{BMWP \text{ score observed by monitoring}}{BMWP \text{ score predicted by RIVPACS}}$$

(d) SWQOs – Statutory Water Quality Objectives. These define use-related standards as targets to be aimed for to protect specific river uses, in relation to local needs.

SAQ 4

(a) Turbidity depends on the fineness, colour and shape of the suspended particles, while the concentration of suspended solids is a measure of the mass of solids present in a volume of water. Thus there is not usually a direct link between turbidity and suspended solids.

(b) The water could be acceptable provided it met all the other requirements, as no standards have been set for turbidity in water to be abstracted for use as drinking water. The degree of turbidity would affect the treatment required.

SAQ 5

For arsenic, the limit will be 50 μg l^{-1}, which is also the limit for salmonid fish.

For iron in solution, the limit will be 2000 μg l^{-1} or 2 mg l^{-1} for A2 waters.

SAQ 6

(a) No, it would not meet the standards. The maximum standard set by the EU Drinking Water Directive is 50 mg l^{-1} expressed as NO_3^- which is equivalent to 11.3 mg l^{-1} expressed as N.

(b) Only the conductivity, and copper and phenol concentrations are outside the limits for A2 treatment. The conductivity and copper concentrations exceed the guide values but are not considered injurious to health. The phenol concentration is, however, above the imperative value for A2 treatment but not significantly so. In general, an A2 treatment would suffice.

In many ways it is really a matter of opinion whether any particular process used will equate to A1, A2 or A3 treatment. The Annex to the directive would have to be very large if all possible treatments were to be covered, and it would also be confusing! Thus there has to be a subjective assessment to decide into which category a particular treatment will be placed.

SAQ 7

Comparison has to be made with the limit values given in the Water Supply (Water Quality) Regulations 2000 for England and Wales.

The pH is below the range allowed (6.5–10.0), and would dissolve lead from piping. Colour and turbidity levels are beyond the values allowed (20 Hazen units and 4 NTU, respectively).

Chloride is also beyond the allowed limit (250 mg l^{-1}).

Iron at 0.3 mg l^{-1} is beyond the 200 μg l^{-1} allowed.

Copper is just at the limit (2 mg l^{-1}), specified in the Drinking Water Directive.

No limit for calcium is specified in the Regulations.

Lead will exceed the limit (25 μg l^{-1} from 2003–2013).

Arsenic is also beyond the limit of 10 μg l^{-1}.

All in all, it's a poor quality supply!

SAQ 8

Coagulation is the precipitation of a colloid from a solution, brought about by the addition of a coagulant to the water in question.

Flocculation, on the other hand, is the bringing together of finely divided matter in order that flocs are formed.

SAQ 9

Pollutant	Origins/examples	Method of removal
gross solids	branches, twigs, etc.	coarse and fine screens
turbidity	suspended matter, e.g. silt, algae	sedimentation; filtration; flotation; (possible pretreatment using microstrainers, or coagulation and flocculation)
colour	organic substances from decaying plant material	chemical coagulation; Sirofloc process; ozonation
iron	leached from soil	simple aeration; chemical coagulation; biological treatment in filters
manganese	leached from soil	addition of oxidising agent at pH $\geqslant$ 9; biological treatment in filters
aluminium	leached from soil	chemical coagulation

SAQ 10

(a) Trace organics such as pesticides and herbicides, volatile organic compounds and trihalomethanes (THMs), are given attention because of the human health risks inherent in them.

Some trace organics can promote bacterial growth in the distribution system and cause a deterioration in water quality.

(b)

Compounds	Removal system
pesticides and herbicides	granular activated carbon (GAC); biological activated carbon (GAC + ozone or hydrogen peroxide)
volatile organic compounds	GAC; air stripping
trihalomethanes	GAC

It is better if THMs are not allowed to form in the first place. This can be achieved by:

1   removing any THM precursors by chemical coagulation prior to disinfection with chlorine;

2   using activated carbon to remove any THM precursors;

3   controlling the chlorine dose to prevent THM formation;

4   using disinfectants other than chlorine.

(c) The limit is 0.1 μg l^{-1} except for aldrin, dieldrin, heptachlor and heptachlor epoxide, whose limit is 0.030 μg l^{-1}.

SAQ 11

(a) Membranes

(b) Ozone (O_3) and chlorine dioxide (ClO_2).

SAQ 12

(a) Blue-green algae are a toxic species of algae, some of which are capable of fixing nitrogen. They are usually prevalent in the summer months, at times of high solar intensity.

(b) Breakpoint chlorination is chlorination of water until the chlorine demand of all the pollutants is satisfied. Ammonia can be removed by this method, being converted to nitrogen gas:

$$2NH_3 + 3HOCl \longrightarrow N_2 + 3H_2O + 3HCl$$

(c) A biological fluidised bed consists of small solid granular material which is fluidised by the upward flow of effluent. The solid material offers a large surface area on which micro-organisms can grow and be utilised to treat the effluent.

SAQ 13

(c) grit removal; (i) reed beds; (p) grass plots.

SAQ 14

The main purpose of the directive is to prevent further deterioration of the quality of both surface and groundwaters. Achieving and maintaining 'good' status is constantly referred to and is described in the wide-ranging lists of parametric standards and conditions to satisfy ecological quality. Much of the ecological requirement applies directly to surface waters, but the quality and quantity of groundwaters must be given due consideration, particularly those that act as sources of supply for drinking water resources and/or replenishing wetlands, where changes in characteristics could be highly damaging to the ecological balance of the flora and fauna of these sensitive areas. The directive targets the management of all aspects of water use, such as conservation, leisure, recreational, fisheries, flood alleviation, horizon planning for the future and industrial needs to meet socio-economic targets.

In addition to providing a framework of environmental objectives that will prevent further deterioration of water resources, the directive requires the entire water quality and ecological status of all water resources to be progressively improved. Alongside this aim, Member States will be tasked to promote sustainable water consumption and use by communities and communal organisations.

SAQ 15

The three significant changes to the current approach to managing the water environment throughout the European Community are as follows:

1   river basin management planning;

2   setting environmental objectives using both ecological factors and chemical characteristics to protect and enhance aquatic and linked terrestrial ecosystems;

3   public consultation and the integrated active involvement of stakeholders in the planning process.

SAQ 16

Prior to the directive coming into force, the chemical quality of surface and groundwaters was the focus of attention, with the main classes of parameters being:

■ physico-chemical characteristics such as pH value, colour, turbidity, dissolved oxygen and electrical conductivity;

■ nutrients – phosphates, nitrogenous and ammoniacal compounds;

■ metals – iron, manganese, copper, nickel;

■ 'exotic' organic chemicals – detergents, herbicides, pesticides.

The directive will in future require ecological classifications to be developed and used as the main artefact of quality being supported by chemical quality. The development of hydromorphological indicators to categorise the biological life (flora and fauna) of surface waters will be challenging, so that similar indicators for similar types of waters throughout Member States will be used. Particular emphasis will be given to water bodies that are or have been subject to human activities, hence the need to create reference standards for comparative purposes.

SAQ 17

A fundamental objective of monitoring is to assess, and help to quantify, risks to water quality and the use of particular water resources.

Surveillance monitoring is designed to assess which water bodies may be at risk of failing to meet the objectives of the directive. Such monitoring will be capable of detecting long-term trends, but owing to the monitoring frequency, which may vary from monthly to six-monthly assessments, episodes of short-term duration that may impair water quality are likely to be missed by chemical analysis alone. The allied use of biota may indicate some change to the ecology of the water system, so this will then invoke implementation of an investigative programme to attempt to quantify the impact and establish the cause(s) of biotic change.

Operational monitoring is directed towards those water bodies that are likely to be most at risk of failing to meet specified environmental objectives. These programmes will include references to *all* types of discharges and drainage re-entering the water body.

Protected area monitoring programmes will be designed to detect any changes in overall water quality to special Areas of Conservation, where sensitively balanced habitats have been identified or are being developed to protect certain species. Programmes for this purpose will be in place before 2006. Drinking water resources, particularly for public supplies, are included in this category of monitoring so as to detect as soon as possible changes in water quality characteristics and any potential harmful substances likely to prejudice drinking water quality and public health.

Groundwater monitoring programmes are dual purpose, providing both water quality and volumetric data. The outputs from this type of programme will be used to create models (particularly for those aquifers where water quality, rates of abstraction and replenishment have to be fully established and understood) to protect associated usage as water resources for public health, and commercial uses, and for the protection of terrestrial ecosystems.

Some monitoring programmes will include the use of 'indicators', that is, surrogate measurements to assess individually the impact of some pollutants. The use of such indicators is likely to increase speed of response to determining results. The design, frequency and range of analytical techniques to be used will inevitably incorporate the appropriate use of indicators.

SAQ 18

The colour would probably have been reduced through oxidation of the colour-conferring compounds, which are usually organic. Turbidity reduction would have come about with settlement of suspended solids. Ammoniacal nitrogen and BOD would have been reduced through microbial degradation and consumption.

Coliforms would have been reduced in number through predation by other organisms and to some extent destruction by UV radiation from sunlight.

SAQ 19

The screenings will consist mainly of leaves, branches and algae, and these are usually sent to landfill. The screenings could also be sent for composting.

SAQ 20

From Table 6.2 of the Set Book (p.79), a cationic polyelectrolyte would be the best coagulant. Aluminium and ferric salts would require the addition of alkalinity, which could make the process more complex and costly.

SAQ 21

$$\text{Flow of } 1000 \text{ m}^3 \text{ d}^{-1} = \frac{1000 \times 10^3}{24 \times 60} \frac{1}{\text{min}}$$

$$= 694.44 \text{ l min}^{-1}$$

If the dose of Al is 1 mg l^{-1}, this means a dosage of 694.44 mg min^{-1}.

Now 8% Al_3O_3 means that in 100 g of solution there are 8 g of Al_2O_3.

The formula weight of Al_2O_3 is $(27 \times 2 + 16 \times 3) = 102$

$$\text{Proportion of Al in } Al_2O_3 = \frac{54}{102} = 0.529$$

$\therefore$ an 8% solution of Al_2O_3 will have $0.529 \times 8 = 4.232$ g of Al per 100 g of solution.

So, in 100 g of 8% Al_2O_3 solution there will be 4.232 g of Al.

We need 694.4 mg Al per minute.

This means we need

$$\frac{100 \text{ g}}{4.232 \text{ g}} \times 694.44 \text{ mg of 8% } Al_2O_3 \text{ solution}$$

$$= 16409.263 \text{ mg}$$
$$= 16.409 \text{ g}$$

The specific gravity of the 8% Al_2O_3 solution is given as 1.3.

This means its density is 1300 kg m^{-3} or 1300 g l^{-1}.

If we need, each minute, 16.409 g of the 8% solution of Al_2O_3, this means we need a volume of

$$\frac{16.409 \text{ g}}{1300 \text{ g l}^{-1}} = 0.0126 \text{ l}$$

$$= 12.6 \text{ ml}$$

So, the flowrate of Al_2O_3 will be 12.6 ml per minute.

SAQ 22

When used as a primary flocculant, the polyelectrolyte is likely to have a lower molecular weight and a higher charge density than when used as a flocculant aid.

SAQ 23

The amount of colloidal matter removed in 1 day

$$= 1000 \times 10^3 \; 1 \times 7 \frac{mg}{l}$$

$$= 7 \times 10^6 \; mg$$

$$= 7 \; kg$$

The amount of hydroxide floc produced can be calculated

1 mole of Al produces 1 mole of $Al(OH)_3$

i.e. 27 g of Al produces $(27 + 3(16 + 1))$ g of $Al(OH)_3$

27 g of Al produces 78 g $Al(OH)_3$

In SAQ 21, the dosage rate was 1 mg l^{-1}. Since the flow was 1000 m^3 d^{-1}, the amount of Al needed will be 1000 g or 1 kg.

$$\text{This will produce } \frac{78}{27} \times 1 \; kg = 2.89 \; kg \; Al(OH)_3 \; floc.$$

Thus the total weight of solids produced $= 7 + 2.89 = 9.89$ kg

If the solids content of the sludge is 2.5% then the mass of sludge

$$= \frac{9.89}{2.5} \times 100$$

$$= 395.6 \; kg$$

If the density of the sludge is 1100 kg m^{-3}, this would mean a sludge volume of

$$\frac{395.6 \; kg}{1100 \; kg \; m^{-3}} = 0.3596 \; m^3$$

$$= 359.6 \; litres$$

SAQ 24

(a) Retention time $= 650/(25 \times 10^6/1000)$ m^3 m^{-3} d^{-1}

    $= 0.026$ d

    $= 0.026 \times 60 \times 60 \times 24$ s

    $= 2246$ s

(b) Thus $GT = 25 \times 2249$

    $= 56.2 \times 10^3$

(c)

$$G = \left(\frac{P}{\eta V} \right)^{\frac{1}{2}}$$

Thus

$$25 = \left(P/1.15 \times 10^{-3} \times 650 \right)^{\frac{1}{2}}$$

Squaring both sides, and taking P on its own

$$P = 25^2 \times 1.15 \times 10^{-3} \times 650$$

$$= 467 \text{ W}$$

SAQ 25

$$P = Q\rho gh$$

$$h = \frac{P}{Q\rho g}$$

$$= 1000/(0.53 \times 9.80 \times 10^3)$$

$$= 0.2 \text{ m}$$

The extra head loss round baffles h is

$$h = \frac{nv_1^2 + (n-1)v_2^2}{2g}$$

Thus,

$$0.2 = (n \times 0.3^2 + (n-1)0.45^2)/(2 \times 9.8)$$

$$= (0.09n + (n-1) \times 0.2)/(2 \times 9.8)$$

$$= (0.09n + 0.2n - 0.2)/(2 \times 9.8)$$

Rearranging the expression,

$$0.2 \times 2 \times 9.8 = 0.09n + 0.2n - 0.2$$

$$0.29n = 4.12$$

$$n = 14.21 \text{ (say, 14)}$$

Number of baffles $= n - 1$

Thus, 13 baffles are present.

SAQ 26

$$v_{s2} = g(s-1)\,d^2/18v \text{ m s}^{-1}$$

$$= 9.8(2.7 - 1.0)(3.5 \times 10^{-5})^2/18 \times 10^{-6} \text{ m s}^{-1}$$

$$= 11.3 \times 10^{-4} \text{ m s}^{-1}$$

$$= 1.13 \times 10^{-3} \text{ m s}^{-1}$$

As already stated, particles with a settling velocity $v_s \leqslant Q/A$ will be removed in the same proportion that their velocity bears to Q/A. Thus,

$$\frac{v_{s2}}{v_{s1}} = 1.13 \times 10^{-3}/2.50 \times 10^{-3}$$

$$= 0.452 \text{ or } 45.2\%$$

SAQ 27

$$v_s = \frac{Q}{A}$$

$$1.8 \times 10^{-3} = 350 \times 10^{-3}/A$$

$$A = 350 \times 10^{-3}/1.8 \times 10^{-3}$$
$$= 194 \text{ m}^2 \text{ (say, 200)}$$

A safety factor will have to be included in the design. This could range from 2 to 3 times the theoretical figure; therefore, the possible area would be 400–600 m^2.

SAQ 28

circumference $= 2\pi r$

$\therefore$ $94.2 = 2 \times 3.14 \times r$

$r = 15$ m

$$\text{total tank surface area } \pi r^2 = 3.14 \times 15^2$$
$$= 706.5 \text{ m}^2$$

$$\text{area of flocculation zone} = 3.14 \times 5^2$$
$$= 78.5 \text{ m}^2$$

$$\text{surface area of outer chamber} = 706.5 - 78.5$$
$$= 628 \text{ m}^2$$

$$\text{surface loading rate} = 600 \text{ l s}^{-1}/628 \text{ m}^2$$
$$= 0.96 \text{ l s}^{-1} \text{ m}^2$$

SAQ 29

(a) Tube settlers are more rigid than plates, and are therefore not easily deflected under the weight of sludge. They inhibit mass movement of water due to wind or temperature effects.

(b) • Smaller area required.

• Lower retention times required.

• Good removal of algae and light flocs.

• Lower water content of the sludge removed.

• Quick start-up.

SAQ 30

Using Table 8.1 in the Set Book, the best option comes out to be the hopper-bottomed upward flow clarifier.

	Effectiveness with algae	Effectiveness on small works	Ability to withstand sudden changes in water quality	Performance with unskilled operators	Low maintenance requirements	Total score
Conventional horizontal flow	+ +	+ +	+ + + +	+ + + +	+ + +	15
Two level horizontal flow	+ +	+	+ + + +	+ + +	+ +	12
Radial horizontal flow	+ +	+ +	+ + +	+ + +	+ + +	13
Plate or tube settlers	+	+ + + +	+	+	+ + +	10
Hopper-bottomed upward flow	+ + +	+ + + +	+ + +	+ +	+ + + +	16
Flat-bottomed upward flow	+ + +	+ + +	+ +	+ + +	+ + +	14
Dissolved air flotation	+ + + +	+ + +	+	+ +	+ +	12

Note: Scale of + to + + + +, low to high benefit.

SAQ 31

(a) If the weir has a high loading per unit length, settled sludge may be carried over. Increasing the weir length helps to overcome this, and this can be done by having notches cut in the weir, or by using a trough with water entering along both sides.

(b) The Accelator can be circular and is divided into inner and outer zones, the outer zone being cone shaped. Coagulation takes place in the inner zone and the treated water flows upwards in the outer zone. The settled sludge in the outer zone is returned to the inner zone for reuse and this is similar to what happens in the activated sludge process in sewage treatment.

SAQ 32

The uniformity coefficient (UC) is the ratio of the sieve size which allows 60% of the sand to pass to that sieve size which only allows 10% to pass.

From the results given,

sieve size/mm	0.30	0.42	0.60	0.84	1.00
sand passing/%	2	10	60	70	80

therefore,

$$UC = 0.60 / 0.42$$
$$= 1.43$$

The sand would be suitable.

SAQ 33

In slow sand filters the filtering action takes place close to the sand surface, i.e. the *Schmutzdecke*, while for rapid sand filters the action takes place mainly deep in the filter, the surface film only making a small contribution to the filtering action.

Rapid gravity sand filters can be cleaned by backwashing, while slow sand filters require the surface sand to be removed manually for washing (a process known as skimming).

SAQ 34

1   A sand bed in which filtration takes place.

2   A support for the sand bed, e.g. gravel, porous plate.

3   An underdrain system to carry away filtered water and admit backwash water and air for agitation, if required.

4   An inlet for water.

5   An outlet for washwater.

6   Means of controlling the flow through the filter.

SAQ 35

Time to backwash each filter:

$$6 \times 60 = 360 \text{ s}$$

Volume of water for backwash in each filter $= 360 \text{ s} \times 15 \times 10^{-3} \text{ m s}^{-1} \times 75 \text{ m}^2$
$$= 405 \text{ m}^3$$

Thus for 4 filters, water required

$$= 4 \times 405 \text{ m}^3$$

$$= 1620 \text{ m}^3$$

SAQ 36

A   True. Above pH 7.5, hypochlorous acid is 50% dissociated. Below pH 7.0, dissociation is 10%.

B   False. The disinfecting agent is hypochlorous acid ($HOCl$).

C   False. Hypochlorous acid is referred to as free available chlorine.

D   True.

E   False. At breakpoint most chloramines will have reacted to form nitrogen, nitrous oxide and other products.

F   False. Chlorine dioxide is not stable and the possible formation of THMs has not been fully investigated.

G   False. The excess chlorine must be removed to prevent corrosion and taste problems.

SAQ 37

For 99.9% kill,

$$t^2 = \frac{2}{10^{-2}} \ln \left(\frac{100}{0.1} \right)$$

$$= 200 \ln 1000$$

$$= 200 \times 6.908$$

$$= 1381.6$$

$$t = 37.2 \text{ s}$$

SAQ 38

$$\ln N_t/N_0 = -kt$$

Thus,

$$\ln 0.1/100 = -3 \times 10^{-2}t$$

$$t = \ln 0.001 \times \frac{1}{\left(-3 \times 10^{-2}\right)}$$

$$= -6.908 \times \frac{1}{\left(-3 \times 10^{-2}\right)}$$

$$= 230.3 \text{ s}$$

SAQ 39

Treatment of industrial wastewater is more difficult due to its:

- greater variability in flowrate and composition;
- likelihood of containing chemicals that are toxic (to micro-organisms at the treatment works), corrosive or of extreme pH, or a combination of any of these;
- likelihood of causing shock loading at treatment works, because of the erratic discharge of specific chemicals and waste.

SAQ 40

Let allowed concentration of Cr in trade effluent be x mg l^{-1}.

Total Cr in trade effluent $= (1.4 \times 10^6)(x)$ mg

70% is removed in primary sedimentation, so 30% goes for secondary treatment.

Thus $(0.3)(1.4 \times 10^6)(x)$ mg goes for secondary treatment.

20% of this appears in final effluent, i.e.

$$(0.2)(0.3)(1.4 \times 10^6)(x) \text{ mg}$$

In terms of concentration,

$$= \frac{(0.2)(0.3)(1.4 \times 10^6)(x)}{60 \times 10^6} \frac{\text{mg}}{\text{l}}$$

Cr standard for abstraction is 0.05 mg l^{-1}

Since effluent from the sewage treatment works is diluted fourfold, the maximum concentration allowed from the works is

$$(0.05)(4) = 0.20 \text{ mg l}^{-1}$$

$$0.20 = \frac{(0.2)(0.3)(1.4 \times 10^6)(x)}{60 \times 10^6}$$

$$x = \frac{60}{(0.3)(1.4)}$$

$$= 142.86 \text{ mg l}^{-1}$$

SAQ 41

Some advantages of using the COD test are:

1   The COD test achieves a higher degree of oxidation of organic matter than the 5-day BOD test, thus reducing the risk of error.

2   The COD test is simpler, quicker and is more reproducible than the BOD test.

3   The COD test provides a financial incentive for industry to use materials that are degradable, whereas the BOD test encourages the use of inhibitory or toxic compounds which reduce the BOD value.

4   The COD test gives a measure of the total load on the water system as it includes both oxidisable organic and inorganic material.

5   The BOD test requires the addition of nutrients and seed water and also takes no account of the varying oxidation rates of trade effluents.

Some advantages of using the BOD test are:

1   The prime objective of the charging scheme is to ensure that the trader pays a fair charge for the services rendered in the treatment and disposal of the trader's effluent. The size, capital cost and running costs of a biological treatment plant are more closely a function of the BOD load than the COD load.

2   The ratio of COD:BOD in industrial wastes varies from 1:1 to 7:1 as industrial wastes are often not biodegradable. Thus the COD could lead to unfair charging.

3   Pretreatment of industrial wastes usually removes more BOD than COD, hence if the charge is based on COD the discharger will not get the full benefit of the work s/he does in pretreatment.

4   The reproducibility of the BOD test has been criticised. However, within any one laboratory the two tests are comparable, and if the industrial effluent has a high chloride concentration then the BOD test tends to be more reproducible.

5   It can be maintained that the BOD test encourages the use of toxic or inhibitory compounds. It must be remembered that these compounds are controlled by the consent conditions, so it can be argued that the use of the BOD test would help in routine analysis to detect the presence of inhibitory compounds in the effluent.

These factors show the arguments put forward by those who favour each of the tests. In general the COD test tends to be more acceptable because less analytical skill is required in carrying it out and also because it requires only hours instead of days to complete. This is why it is accepted by both the CBI and the water and sewerage companies.

SAQ 42

(a) (i)

$$\text{Cost per m}^3 = R + V + \left(\frac{O_t}{O_s}\right)B + \left(\frac{S_t}{S_s}\right)S \tag{S1}$$

where

$R = 13.17\text{p m}^{-3}$

$V = 12.74\text{p m}^{-3}$

$O_t = 1500 \text{ mg l}^{-1}$

$O_s = 542.8 \text{ mg l}^{-1}$

$B = 18.55\text{p m}^{-3}$

$S_t = 250 \text{ mg l}^{-1}$

$S_s = 347.6 \text{ mg l}^{-1}$

$S = 11.97\text{p m}^{-3}$.

Substituting into Equation S1,

$$\text{Cost per m}^3 = 13.17 + 12.74 + \left(\frac{1500}{542.8}\right)18.55 + \left(\frac{250}{347.6}\right)11.97$$

$$= 85.78\text{p}$$

Volume discharged per day $= 300 \text{ m}^3$

Therefore

$$\text{Total daily charge (£)} = 300 \times \frac{85.78}{100}$$

$$= £257.34$$

(ii) Volume of discharge after cooling water removed:

$$= \frac{70}{100} \times 300$$

$$= 210 \text{ m}^3 \text{ per day}$$

The COD and the suspended solids figures will increase if the pollutant load remains the same. Say the new COD concentration is x; then

$$(x)(210) = (1500)(300)$$

$$x = \frac{(1500)(300)}{210}$$

$$= 2142.86 \text{ mg l}^{-1} \text{ (say, 2143 mg l}^{-1}\text{)}$$

Similarly for suspended solids, if the new concentration is y, then

$$(y)(210) = (250)(300)$$

$$y = \frac{(250)(300)}{210}$$

$$= 357.14 \text{ mg l}^{-1} \text{ (say, 357 mg l}^{-1}\text{)}$$

$$\text{New cost per m}^3 = 13.17 + 12.74 + \left(\frac{2143}{542.8}\right)18.55 + \left(\frac{357}{347.6}\right)11.97$$

$$= 111.44\text{p}$$

$$\text{Total daily charge (£)} = 210 \times \frac{111.44}{100}$$
$$= £234.02$$

(iii)

$$\text{Cost per m}^3 = 13.17 + 12.74 + \left(\frac{700}{542.8}\right)18.55 + \left(\frac{50}{347.6}\right)11.97$$
$$= 51.55\text{p}$$

$$\text{Total daily charge (£)} = 210 \times \frac{51.55}{100}$$
$$= £108.26$$

(iv) The charge would be based on R (reception and conveyance costs), and M (marine costs).

$$\text{Total daily charge (£)} = \frac{(13.17 + 9.59)}{100}300$$
$$= £68.28$$

While the treated effluent costs less to discharge, detailed costing of the treatment process should be considered before this option is chosen, as it may make the overall cost higher than direct discharge to sewer. Volume reduction would probably be the most effective means of reducing the cost of disposal.

(b) (i) Here we need to use Equation S1 above with all the parameters fixed except for O_t, which will vary from 1500 mg l^{-1} to 0 mg l^{-1} (say). S_t becomes

$$S_t = \frac{5}{100} \times 250 = 12.5 \text{ mg l}^{-1}$$

since 95% removal of suspended solids is achieved in pretreatment.

$$\text{Cost per m}^3 = R + V + \left(\frac{O_t}{O_s}\right)B + \left(\frac{S_t}{S_s}\right)S$$

where
$R = 13.17\text{p m}^{-3}$
$V = 12.74\text{p m}^{-3}$
$O_t = \text{variable mg l}^{-1}$
$O_s = 542.8 \text{ mg l}^{-1}$
$B = 18.55\text{p m}^{-3}$
$S_t = 12.5 \text{ mg l}^{-1}$
$S_s = 347.6 \text{ mg l}^{-1}$
$S = 11.97\text{p m}^{-3}$

$$\text{Cost per m}^3 = 13.17 + 12.74 + \left(\frac{O_t}{542.8}\right)18.55 + \left(\frac{12.5}{347.6}\right)11.97$$
$$= 26.34 + 0.0342O_t \text{ pence}$$

We can calculate the cost per m^3 for O_t varying from 1500 to 0 mg l^{-1}. Using steps of 100 mg l^{-1}, results as in Table 24 are obtained.

Table 24

O_t/(mg l^{-1})	Cost per m^3 (pence)
1500	77.64
1400	74.22
1300	70.80
1200	67.38
1100	63.96
1000	60.54
900	57.12
800	53.70
700	50.28
600	46.86
500	43.44
400	40.02
300	36.60
200	33.18
100	29.76
0	26.34

A graph of cost per m^3 of effluent discharged versus effluent COD can be generated (Figure 47).

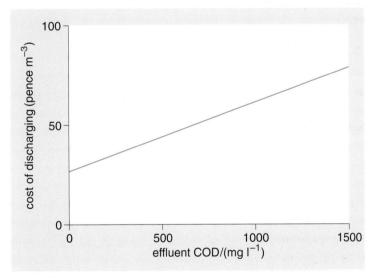

Figure 47

(ii) The cost of discharging the untreated effluent is (from a,i) 85.78 pence per m^3. Half this cost is 42.89 pence per m^3. From the graph in (b,i), pretreatment to bring the level of the COD in the original effluent to about 500 mg l^{-1} is necessary to arrive at this figure.

We can generate a second graph giving greater resolution around the value (Figure 48). By inspection, the required COD value is found to be about 484 mg l^{-1}.

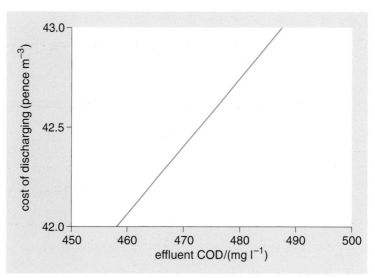

Figure 48

We can verify this by calculation, using the equation

$$\text{Cost per m}^3 = 26.34 + 0.0342\,O_t$$

If we substitute 42.89 pence for the cost per m^3, we get:

$$42.89 = 26.34 + 0.0342\,O_t$$

$$O_t = \frac{42.89 - 26.34}{0.0342}$$

$$= 483.92 \text{ mg l}^{-1}$$

SAQ 43

There are several reasons why waste minimisation is especially critical with regard to water.

■ Potable water has a cost element associated with it. It may seem cheap at the moment but with increasingly stringent regulations as to its quality (as is the case in Europe) it will become more expensive. Hence, water wasted is indeed 'money down the drain'.

■ Used water is usually sent to a treatment works before discharge to a watercourse. Treatment works are designed on the basis of volumetric throughput. If unnecessarily high volumes of water are used and discharged, high capital costs are incurred in setting up and operating large treatment plants.

■ It is easier to treat a concentrated waste and, in the case of an industrial effluent, the opportunity for product recovery (e.g. by membrane filtration) may be presented.

■ Water may be scarce in a given environment. For instance, in the Middle East, an oil refinery may be economical with water use and generate only about 65 litres of process wastewater per barrel of crude refined, while an equivalent refinery in a water-rich area may produce about 135 litres per barrel of crude.

■ Greater awareness of environmental issues among people both in industry and outside it has prompted action to be taken with regard to minimisation of effluent discharges. This has led to waste minimisation through the implementation of better process control and the adoption of cleaner technologies where possible.

SAQ 44

From T210/T237 *Environmental Control and Public Health*, the options shown in Tables 25 and 26 are available.

Table 25   Ways to conserve potable water within the home

Advice	Notes
Fix leaking taps.	A dripping tap wastes about 13 litres of water a day.
Never let the water run while brushing teeth, etc.	Leaving the tap running can waste up to 9 litres a minute.
Install spray nozzles on taps.	These reduce consumption by 10–20%.
Use a pressure cooker instead of a normal saucepan.	This needs less water.
Use water-economical washing machines and dishwashers, and use them only with a full load.	Typically, washing machines consume 65 litres per wash. Requirements under the 1999 Water By-laws limit the water use per cycle in new machines to 35 litres. Some washing machines can 'sense' the size of the washing load, and adjust their water consumption accordingly. Such machines get a well-deserved *Eco-label*!
Have a shower instead of a bath.	On average, a 6-minute shower will use up to 30 litres of water, while a bath will consume 80 litres. (Power showers, however, will consume more water than a bath if used for more than five minutes.)
Use a cistern device such as a bottle or a 'Hippo' (a plastic bag to hold water) to displace a given volume of water.	Reduces the amount of water used in each flush. The volume displaced by the various devices range from 1 litre to 3 litres.
Use new-model WCs which use less water.	Regulations introduced in 1999 for new installations from 2001 reduced the amount of water used to no more than 6 litres per flush. Dual-flush WCs can also reduce consumption, if used correctly.
Use rainwater for toilet flushing and for outside use.	Not all uses need potable quality water.
Reuse greywater (the non-faecally contaminated used water from baths, showers and hand basins) for toilet flushing, after appropriate treatment.	Water from kitchen sinks, washing machines and dishwashers usually contains food particles, grease and oil that are difficult to filter.
Use supply restrictor valves (plumbing fittings that reduce water flows into domestic appliances, taps and showers).	Essentially, the valve works by reducing the diameter of the supply pipe when there is a change in mains pressure, thus giving a constant flow. The result is that even when the tap is fully open the same flow rate of water is maintained. A valve is also available that can detect leaks and shut off supplies automatically.

Table 26   Ways to use less water in gardening

Advice	Notes
Collect rainwater by connecting a water butt to the downflow pipe from the roof.	The water is free, and good for plants.
Water the garden in the evening.	The rate of evaporation is lower at this time.
Use mulch (e.g. chopped bark, grass cuttings, polythene mulching mats)	Reduces the extent of evaporation from the soil around plants.
Use drip-feed irrigation.	Ensures water reaches only the required parts of the garden.
Water the lawn only once a week (twice a week in the hottest weather) with a sprinkler moved around the garden.	Watering more frequently encourages the roots of the grass to seek the surface, and overwatering encourages moss. Spiking the lawn will help the water soak down.
Don't cut the grass too short.	Longer grass shades the soil so that it retains moisture longer, and stays green longer, too.
Use a watering can instead of a sprinkler.	On average, in just one hour, a sprinkler will use as much water as a family of four uses in two days.
If a hose is used, equip it with a trigger.	The water supply can be cut off when it is not needed.
Add compost or organic matter to soil.	Improves water retention.
Build paths of gravel instead of paving slabs.	All the rainwater seeps into the garden.
Grow drought-resistant plants which require less water.	For example, climbing roses, vines, red hot pokers, poppies, cornflowers, lavender, thyme, rosemary, mint and sage. This is known as 'xeriscaping' or landscaping for water conservation.
Don't water trees and shrubs once they are established.	Watering brings the roots to the surface. Established roots will seek out groundwater so don't need watering.
Only water frequently those fruits and vegetables that have edible leaves, and water root crops and tomatoes only when their roots or fruits are swelling.	Watering at other times is not necessary.
Insert sections of downpipe into the soil beside plants.	The water for irrigating them reaches the roots directly.
Reuse dirty water from fish tanks for the garden or houseplants.	It's rich in nitrogen and phosphorus.

SAQ 45

A   False. The air pressure is about half that of atmospheric pressure.

B   True.

C   False. The wood shavings help to create voids to allow air through the mass of solids.

D   True.

E   False. Electricity is also needed to power the rotator for the sewage, and to heat and dry the solids. It also gives power to the electronic control system.

F   True.

SAQ 46

(a) The kinds of pollutants likely to be present are oil and grease, sand and detergent. Gravity separation would remove the sand and some of the oil and grease. The remaining oil and grease could then be removed by flotation, membranes, filtration or centrifugation. Economics and other factors (e.g. ease of maintenance) will determine which option is finally chosen. Chemicals may be needed to de-emulsify the oil. The residual detergent in the treated water will be beneficial when the water is reused.

(b) The residues requiring disposal would be the collected sand, oil and grease, and possibly chemical containers. The sand, oil and grease can be landfilled, while the containers can be returned to the chemical supplier.

SAQ 47

The programme should begin with a waste audit, focused on water usage. To be effective, the audit should have the support of senior management. The following are the main stages of the programme.

1 Implementing the waste audit.

   ● Formation of the audit team using personnel from a range of disciplines.

   ● Acquisition of information on the various processes. This will involve: site tours; drawing process plans and diagrams; quantifying and characterising the waste streams; verifying that drainage drawings represent the actual situation on the ground.

   ● Consideration in detail of the quantity and quality of water used on-site and the effluents generated. Opportunities for reducing the water requirements and the amount of wastewater generated will be identified.

2 Option generation.

   ● Generation of waste minimisation options through: brainstorming sessions; ideas from the workforce.

   ● Prioritisation of options based on technical feasibility, effectiveness and economics.

3 Evaluation.

   ● Evaluation of the waste minimisation opportunities identified in terms of monetary returns, and technical and environmental criteria.

   ● Selection of the waste streams for minimisation.

4 Implementation of waste minimisation measures. The selected options are implemented and the plant database is updated.

5 Continuous monitoring and targeting. Monitoring after implementation of the above measures will reveal their effectiveness and highlight new areas to focus on for further improvement. Targets for further consideration can then be identified.

6 Dissemination of results. This is desirable in order to keep up the momentum in the waste minimisation 'journey'.

SAQ 48

(a) From Figure 14, the volume of effluent treated per day is 96.5 m^3 (from low grade water and spent CIP solutions). So, in a year, the volume of effluent treated will be $96.5 \times 5.5 \times 50 = 26\ 537.5$ m^3.

(b) The payback period would be $(1\ 020\ 000/686\ 000) = 1.49$ years.

SAQ 49

	Cost savings ($£\ year^{-1}$)
Avoidance of need for tankering of 47 800 m^3 effluent off-site	468 500
Reduction in mains water consumption by 44 000 m^3	29 480
Replacement of raw materials with 480 tonnes recycled fibre	17 760
Sub-total of cost savings	**515 740**
Less cost of fees to supplier-operator of new treatment plant ($= 12 \times 22\ 000$)	264 000
Net annual cost saving ($= 515\ 740 - 264\ 000$)	**251 740**

SAQ 50

For a water supply of 110 000 $m^3\ d^{-1}$, the number of check monitoring samples per year will be

$$4 + 3\left(\frac{110\,000}{1000}\right) = 4 + 330 = 334$$

The number of audit monitoring samples per year will be

$$10 + 1\left(\frac{110\,000}{25000}\right) = 10 + 4.4 = 14.4 \quad \text{i.e.}\,15$$

SAQ 51

For both types of analysis care must be taken to ensure that no contamination occurs as a result of the sampling technique. Thus for chemical analysis the clean sample bottle must be well rinsed with the water to be sampled, to avoid contamination from previous samples or bottle-washing solutions. Smple bottles will often have preservatives (e.g. dilute acid to keep metal ions in solution, or caustic soda to stabilise cyanide complexes) to ensure that the parameters of the sample do not change. Appropriate bottles must be used – e.g. if analysing for organic chemicals, ultraclean borosilicate glass bottles must be used. If sampling for lead in water in the home, the first runnings must be collected to check for plumbosolvency.

For bacteriological samples, not only the sample bottle but also the sampling tap must be sterilised. The sample bottle will need a few drops of sodium thiosulphate to neutralise any residual chlorine in the water. The sample should be transported at low temperature.

SAQ 52

A False. Instantaneous breathing rates are compared with historical data by a computer and only after statistical analysis is a decision made as to the change in water quality.

B False. The fish are not fed while 'on-duty' as the metabolic processes related to digestion would affect the monitoring process.

C True.

D False. The samples are compared with historical data of clean river water taken at approximately the same time of day.

E True.

F False. Organelles or parts of organisms can also be utilised.

G True.

SAQ 53

Faecal coliforms indicate that pathogenic bacteria could be present. This could be the result of inadequate disinfection at the treatment works or ingress of contaminating matter into the distribution system.

The first action should be to take a repeat sample from the original sample point and then other samples in the immediate vicinity, such as at adjacent properties. As a precautionary measure, mains water in the locality may be flushed to waste. The records of the disinfection process at the water-works supplying that water will be checked for any irregularities, together with records of any mains work in the area. Other activities by third parties (e.g. fire services) are also checked.

If the other samples, and the repeat samples, are free from faecal coliforms, then no further action is required except to report the incident to the local authority and the health authority.

If the extra samples show that faecal coliforms are still present, action such as increased flushing, adding additional chlorine to that area from a service reservoir, and increased monitoring is necessary. The CCDC and the relevant EHO should be informed via the action plans set up, and consideration given to issuing customers with advice to boil water. The Drinking Water Inspectorate (DWI) will also have to be informed. Further action will depend on the severity of the incident.

SAQ 54

The river will have the following characteristics:

- a large variation in flow
- variations in water quality – high suspended solids content may be possible during high flow or pollution incidents caused by the intensive dairy farming; this farming may also place an organic loading on the river
- a possible high bacterial count from the intensive dairy farming
- possible contamination from the leaching from the soil of the high concentration of nutrients used in farming
- reduction of the dissolved oxygen content in the river due to oxidation of the iron and manganese from the mine workings, and discoloration due to iron and manganese in suspension.

From the information given (which does not include any analytical details, or information on river flow, volume of water to be abstracted, population to be served, etc.), it is not possible to state with certainty what sort of treatment will be required. It is possible, however, to give a general indication of the most likely treatment by referring to Table 3.1 of the Set Book, and also by considering the information given throughout the Set Book.

The following is a possible treatment option (summarised in Figure 49).

After abstraction and passing through coarse screens, the raw water would need to be stored to maintain the supply when the abstraction is limited due to the poor quality of the water source (high flow or pollution). The Set Book (Chapter 4) suggests storage of 7 days. This storage will help to reduce suspended solids of silt, iron and manganese, and also reduce the likely bacterial count which may be present as a result of the intensive dairy farming. It is worth noting that if a river regulation scheme is introduced the problems of the uneven flow would be overcome.

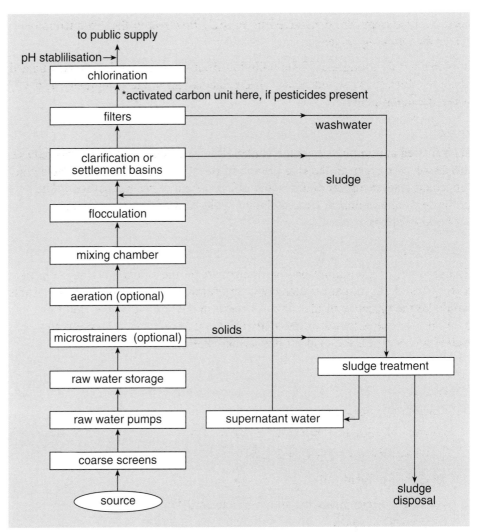

Figure 49   Summary of treatment processes for the water source in SAQ 54

Algal growth is likely to occur periodically due to the presence of nutrients from the farming activity, and copper sulphate may sometimes have to be added to the water during storage. Microstraining could be an optional extra as it would help to reduce any algal growth or suspended matter passing on to the next stage of treatment. Aeration may be required as the dissolved oxygen in the water may be low as a result of the oxidation of the iron and manganese, and also the biological oxidation of any organic farming pollution. It is unlikely that any prechlorination will be required as no industrial or domestic sewage has apparently entered the river. The water is unlikely to be highly coloured as it has not been derived from a 'peaty' source, but it may on occasions be turbid.

The remaining treatment will consist of coagulation, flocculation, settlement, rapid filtration and disinfection. This format of treatment is required because of the likely presence of suspended matter, turbidity, algal growth and coliform bacteria. If pesticides are present in the outlet from the filter, activated carbon may have to be used.

Turbid water will usually form a good floc and settle rapidly, and for this type of water a wide choice of systems exists. If the flow can be kept constant then settlement tanks using a sludge blanket will operate well. The tank design will depend on the area of land available, and the depth of the tank will depend on the type of ground available. These factors will delineate whether radial or rectangular, cone-shaped or flat-bottomed tanks are used. The water is likely to need A2 type treatment considered in the EU Directive on the Abstraction of Water for Drinking Water.

It is also worth noting that if the pollution from the mine workings were treated at source, then there would be a considerable saving in the complexity of the water treatment plant.

SAQ 55

Both treated and untreated sludge contain biodegradable matter and this reduces the dissolved oxygen in the river. Some of the sludge is not biodegradable and this could lead to high turbidities and affect the entire river ecosystem. The presence of aluminium in treated sludge could cause fish kills and other ecological damage.

SAQ 56

The presence of aluminium or iron compounds in the sludge may be beneficial as they could aid primary settlement at the sewage works. A direct result of this would be the presence of aluminium or iron in the sewage sludge and this could restrict the disposal of the sludge to land. In general, the waterworks sludge will increase the load on the sewage works.

SAQ 57

(a) Given

flowrate of river $= 720$ m^3 h^{-1} $= 0.2$ m^3 s^{-1}

flowrate of effluent $= 90$ m^3 h^{-1} $= 0.025$ m^3 s^{-1}

BOD_5^{20} of river $= 2$ mg l^{-1} $= 2$ g m^{-3}

BOD_5^{20} of effluent $= 30$ mg l^{-1} $= 30$ g m^{-3}

velocity of river flow $= 0.1$ m s^{-1} $= 8640$ m d^{-1}

k_L at 20 °C $= 0.25$ d^{-1}

BOD_5^{20} of river water/effluent mixture

$$= \frac{(2)(0.2) + (30)(0.025)}{0.2 + 0.025}$$

$$= \frac{0.4 + 0.75}{0.225}$$

$$= \frac{1.15}{0.225}$$

$$= 5.11 \text{ mg l}^{-1}$$

$$L_0 = \frac{BOD_5^{20}}{1 - e^{-kt}} = \frac{5.11}{1 - e^{-(0.25)(5)}}$$

$$= 7.16 \text{ mg l}^{-1}$$

$U_0 = 8640$ m d^{-1}

For Al treatment, the BOD_5^{20} of the raw water should be less than 3.0 mg l^{-1}.

This is equivalent to an ultimate oxygen demand of:

$$\frac{3}{1 - e^{(-0.25)(5)}} = 4.20 \text{ mg l}^{-1}$$

i.e. $L = 4.20$ mg l^{-1}

Substituting in equation:

$$L = L_0 \exp\left[-\frac{k_L x}{U_0}\right]$$

$$4.2 = 7.16 \exp\left[-\frac{k_L x}{8640}\right]$$

$$\frac{4.2}{7.16} = \exp\left[-\frac{k_L x}{8640}\right]$$

$$0.587 = \exp\left[-\frac{k_L x}{8640}\right]$$

Taking logs base e of both sides,

$$\ln(0.587) = \frac{-k_L x}{8640}$$

$$-0.533 = \frac{-k_L x}{8640}$$

$$\frac{(0.533)(8640)}{k_L} = x$$

$$\frac{4605.12}{k_L} = x$$

Now $k_T = k_{20}(1.047)^{T-20}$

The k_L value for different temperatures from 6 °C to 14 °C can be calculated.

$T\,(°C)$	$k_L\,/(\mathrm{d}^{-1})$
6	0.131
7	0.138
8	0.144
9	0.151
10	0.158
11	0.165
12	0.173
13	0.181
14	0.190

The (theoretical) safe distance x at which the water intake could be located is given by

$$x = \frac{4605.12}{k_L}$$

For each of the temperatures 6 °C to 14 °C, a value of x can be calculated.

$T(°C)$	$k_L /(d^{-1})$	x /km
6	0.131	35.15
7	0.138	33.37
8	0.144	31.98
9	0.151	30.50
10	0.158	29.15
11	0.165	27.91
12	0.173	26.62
13	0.181	25.44
14	0.190	24.24

(b) The values of x calculated are the theoretically safe distances downstream from which water could be abstracted for the various river water temperatures. To cater for the worst case, the intake should be at least 36 km downstream of the effluent discharge.

The plot of river water temperature against location of the water intake is shown in Figure 50.

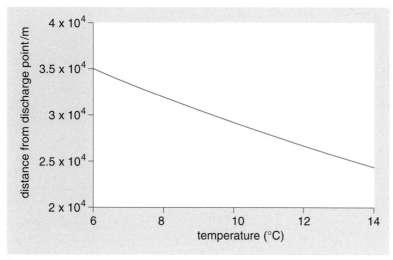

Figure 50

SAQ 58

(a)

$$DO = DO_{sat} - D$$

$$= 9.8 - \left\{ D_0 \exp\left[-\frac{k_a x}{U_0} \right] + \frac{L_0 k_L}{k_a - k_L} \left[\exp\left(-\frac{k_L x}{U_0} \right) - \exp\left(-\frac{k_a x}{U_0} \right) \right] \right\}$$

Substituting,

$D_0 = 8.0$ mg l^{-1}

$k_a = 2.0$ d^{-1}

$U_0 = 0.1$ m s^{-1} = 8640 m d^{-1}

$L_0 = 80$ mg l^{-1}

$k_L = 0.25$ d^{-1}

we can generate the required graph (Figure 51).

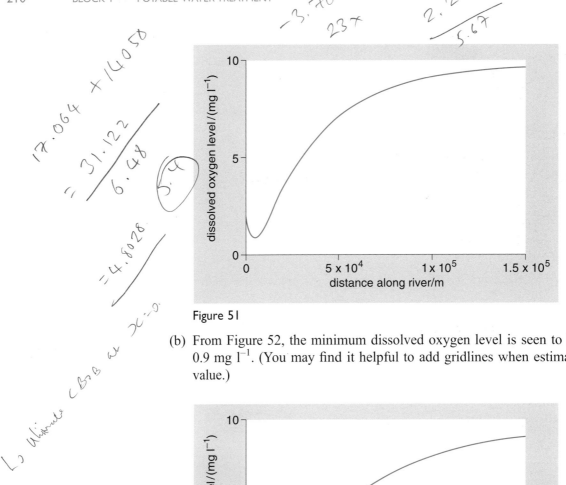

Figure 51

(b) From Figure 52, the minimum dissolved oxygen level is seen to be about 0.9 mg l⁻¹. (You may find it helpful to add gridlines when estimating this value.)

Figure 52

∴ The critical dissolved oxygen deficit = 9.8 − 0.9 = 8.9 mg l⁻¹.
The critical distance is about 4.0 km from the discharge point.

SAQ 59

We need to use the equation

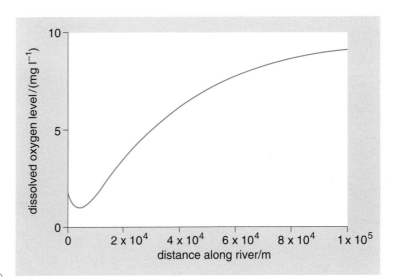

$$D_c = \left[D_0 - \frac{L_0 k_L}{k_a - k_L} \right]\left[\frac{k_a}{k_L}\left(1 - D_0 \frac{(k_a - k_L)}{k_L L_0} \right) \right]^{-\frac{k_a}{k_a - k_L}}$$

$$+ \left[\frac{L_0 k_L}{k_a - k_L} \right]\left[\frac{k_a}{k_L}\left(1 - D_0 \frac{(k_a - k_L)}{k_L L_0} \right) \right]^{-\frac{k_L}{k_a - k_L}}$$

(S2)

Handwritten annotations:

-3.76 $23 \times$ 2.2 $\overline{5.67}$

17.064×14050
$= 31.122$
$\overline{6.48}$
5.4

$= 4.8028$

L_0 ultimate CBOD at $x = 0$

$0777061011 2$

23

$\left[4.66 - \dfrac{52.71 \times 0.11}{0.621 - 0.11} \quad \dfrac{5.7981}{0.511} \quad 11.35 \right]$

$\left[\dfrac{52.71 \times 0.11}{0.621 - 0.11} \right]$

$\dfrac{0.621}{0.11} = 5.645$ $1 - 4.66 \dfrac{(0.621 - 0.11)}{0.11 \times 52.71}$
$1 - 4.107$

$\dfrac{0.511}{5.7981}$

$\dfrac{0.621}{0.621 - 0.11}$

$\dfrac{k_a}{k_a - k_L}$ $\dfrac{k_L}{k_a - k_L}$

$- \dfrac{0.11}{0.621 - 0.11}$

$D_c =$

$D_c = \left(4.66 - 11.35 \right)\left[\left(5.645 \right)\left(-3.107 \right) \right]^{-1.215}$
$+ \left(11.35 \right)\left[\left(5.645 \right)\left(-3.107 \right) \right]^{-0.215}$

$D_c = -6.69 \times \left[-17.54 \right]^{-1.215}$

(a) Determination of D_0:

$$D_0 = \frac{D_{up}Q_{up} + D_e Q_e}{Q_{up} + Q_e}$$

[handwritten: $\dfrac{3.6 + 3.6 + 1.8}{3}$]

[handwritten: $\dfrac{12.6 + 108}{5.4}$]

[handwritten: $\dfrac{120.6}{54}$ 22.333]

River temperature = 20 °C

Saturation concentration of dissolved oxygen at 20 °C:

$= 14.65 - 0.410\ 22(20) + 0.007\ 91(20)^2 - 0.000\ 077\ 74(20)^3$

$= 8.99$ mg l^{-1} *[handwritten: 9.38]*

$D_{up} = 8.99 - 7.0 = 1.99$ mg l^{-1} *[handwritten: $D_{up} = 8.99 - 3.5 = 5.49 mgl^{-1}$]* *[handwritten: $D_{up} = 8.99 - 5.5 = 3.49 mgl^{-1}$]*

$D_e = 8.99 - 0.1 = 8.89$ mg l^{-1} *[handwritten: $D_e = 8.99 - 60$]* *[handwritten: $D_e = 8.99 - 2 = 6.99 mgl^{-1}$]*

$Q_e = 0.4$ m^3 s^{-1} *[handwritten: $Q_e = 1.8 n^3 s^{-1}$]*

$Q_{up} = 2.0$ m^3 s^{-1} *[handwritten: $Q_{up} = 3.6 n^3 s^{-1}$]*

Thus, *[handwritten: Thus:]*

$$D_0 = \frac{(1.99)(2.0) + (8.89)(0.4)}{2.0 + 0.4}$$

$= 3.14$ mg l^{-1}

[handwritten: $D_o = \dfrac{(3.49)(3.61) + (6.99)(1.8)}{3.61 + 1.8}$]

[handwritten: $= \dfrac{12.6 + 12.6}{5.41}$ $= \boxed{4.66 mgl^{-1}}$ D_o]

[handwritten: $\dfrac{5.67}{2.2} = 1.213636$]

(b) Determination of L_0:

[handwritten: R.BOD] BOD$_5^{20}$ of river water/effluent mixture:

$$= \frac{(1.5)(2.0) + (450)(0.4)}{2.0 + 0.4}$$ *[handwritten labels: River(flow), Effluent, flow, flow]*

$= 76.25$ mg l^{-1}

[handwritten: $= \dfrac{(3.5)(3.51) + (60)(1.8)}{3.61 + 1.8}$]

[handwritten: $\dfrac{12.63 + 108}{5.41}$]

[handwritten: $\boxed{L_0 = 22.3 mgl^{-1}}$]

$$L_0 = \frac{BOD_5^{20}}{1 - e^{-kt}} = \frac{76.25}{1 - e^{-(0.3)(5)}}$$

$= 98.15$ mg l^{-1}

[handwritten: $L_0 = \dfrac{22.3}{1 - e^{-(0.11)(5)}}$]

[handwritten: $\boxed{= 52.712 mgl^{-1}}$ L_0]

(c) Determination of k_a:

depth $(H) = 0.7$ m *[handwritten: Depth = 2.0m]*

velocity $(U_0) = 0.2$ m s^{-1} *[handwritten: velocity = 0.2 ms^{-1}]*

From Figure 25, k_a is best estimated by the O'Connor–Dobbins correlation.

[handwritten: $v = \dfrac{Q}{A} = v$ $\dfrac{5.4}{13.5 \times 2}$]

[handwritten: $\dfrac{2.236067}{0.5 \times 5}$]

$$k_a = 3.93\frac{U_0^{0.5}}{H^{1.5}}$$ *[handwritten: $k_a = 3.93 \dfrac{0.2^{0.5}}{2^{1.5}}$]*

$= 3.93(0.764)$ *[handwritten: $= 3.93(0.1581)$]*

$= 3.0$ d^{-1} *[handwritten: $k_a = 0.621 d^{-1}$]*

We now have

$D_0 = 3.14$ mg l^{-1} *[handwritten: $D_0 = 4.66$]*

$L_0 = 98.15$ mg l^{-1} *[handwritten: $L_0 = 52.71$]*

$k_L = 0.3$ d^{-1} *[handwritten: $k_L = 0.11 d^{-1}$ *]*

$k_a = 3.0$ d^{-1} *[handwritten: $k_a = 0.621 d^{-1}$]*

[handwritten: $\dfrac{0.4472135959}{2.82842712} = 0.1$ ✓]

(d) Determination of D_c:

Substituting the values for D_0, L_0, k_L and k_a in Equation S2 we obtain:

$D_c = 7.89$ mg l^{-1}

[handwritten: 0.62138]

(e) Determination of x_c:

$x_c = U_0 t_c$

$U_0 = 0.2$ m s$^{-1} = 17280$ m d^{-1}

$$t_c = \frac{1}{k_a - k_L} \ln\left[\frac{k_a}{k_L}\left(1 - D_0 \frac{k_a - k_L}{k_L L_0}\right)\right]$$

$k_a = 3.0$ d^{-1}

$k_L = 0.3$ d^{-1}

$D_0 = 3.14$ mg l^{-1}

$L_0 = 98.15$ mg l^{-1}

Then,

$$t_c = \frac{1}{3.0 - 0.3} \ln\left[\frac{3.0}{0.3}\left(1 - 3.14 \frac{(3.0 - 0.3)}{(0.3)(98.15)}\right)\right]$$

$$= \frac{1}{2.7} \ln\left[10(0.712)\right]$$

$$= 0.73 \text{ d}$$

$$x_c = (17280)(0.73) = 12\,614.4 \text{ m}$$

$$= 12.61 \text{ km}$$

The critical dissolved oxygen deficit is thus 7.89 mg l^{-1} and it occurs 12.61 km after the point of entry of the slurry into the river.

SAQ 60

(a) The concentration profile for total-N is given by the equation:

$$\text{TN} = TN_0 + \frac{\Delta Q x}{Q}(TN_r - TN_0) \tag{S3}$$

Given that

$TN_0 = 0.3$ mg l^{-1}

$$\Delta Q = \frac{5}{15} = 0.33 \text{ m}^3 \text{ s}^{-1} \text{ km}^{-1}$$

$Q = 10 + 0.33x$ m^3 s^{-1}

$TN_r = 1.5$ mg l^{-1}

Substituting in Equation S3,

$$\text{TN} = 0.3 + \frac{0.33x}{10 + 0.33x}(1.5 - 0.3)$$

$$= 0.3 + \frac{0.396x}{10 + 0.33x}$$

A plot of TN versus distance is shown in Figure 53.

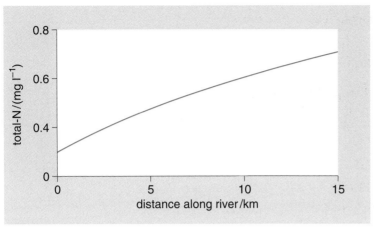

Figure 53

The concentration of total-P in the river is given by the equation

$$TP = TP_0 + \frac{\Delta Qx}{Q}(TP_r - TP_0)$$ (S4)

Given

$TP_0 = 0.03$ mg l^{-1}

$\Delta Q = 0.33$ m^3 s^{-1} km^{-1}

$Q = 10 + 0.33x$ m^3 s^{-1}

$TP_r = 0.25$ mg l^{-1}

Substituting in Equation S4

$$TP = 0.03 + \frac{0.33x}{10 + 0.33x}(0.25 - 0.03)$$

$$= 0.03 + \frac{0.0726x}{10 + 0.33x}$$

Again, a plot of TP versus distance can be generated (Figure 54).

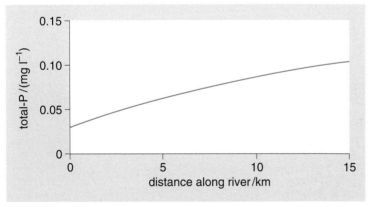

Figure 54

(b) The plot of the ratio TN:TP can be generated as in Figure 55.

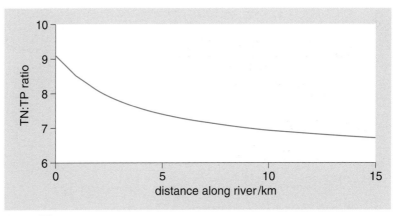

Figure 55

The point at which the TN:TP ratio is 7.5:1, is about 4.0 km along the river. After this point, the deficiency in N limits growth of the algae.

SAQ 61

(a) For A3 treatment (from Appendix 1), the total coliform limit is 50 000 (MPN) per 100 ml.

The equation relating the total coliforms at the water intake (TC) to that at the point in the river where the treated effluent enters (TC_0) is:

$$TC = TC_0 \ \exp\left[-\frac{k_{tc}x}{U_0} \right] \tag{S5}$$

Given

$TC = 50\ 000$ per 100 ml

$k_{TC} = 0.70\ \mathrm{d}^{-1}$

$x = 15\ \mathrm{km} = 15\ 000\ \mathrm{m}$

$U_0 = 1\ \mathrm{m\ s}^{-1} = 86\ 400\ \mathrm{m\ d}^{-1}$

Substituting in Equation S5,

$$50\,000 = TC_0 \ \exp\left[-\frac{(0.70)(15\,000)}{86\,400} \right]$$

$$50\,000 = TC_0 \ \exp[-0.1215]$$

$$TC_0 = \frac{50\,000}{\exp[-0.1215]}$$

$$= 56\,459.47 \ (\text{say, } 56\,460 \text{ per } 100 \text{ ml})$$

TC_0 can be related back to the total coliforms in the river and those in the treated effluent by the equation:

$$TC_0 = \frac{(TC_{up})(Q_{up}) + (TC_e)(Q_e)}{Q_{up} + Q_e} \tag{S6}$$

Substituting

$TC_0 = 56\ 460$ per 100 ml

$TC_{up} = 1000$ per 100 ml

$Q_{up} = 200\ \mathrm{m}^3\ \mathrm{h}^{-1}$

$TC_e = $ unknown

$Q_e = 50\ \mathrm{m}^3\ \mathrm{h}^{-1}$,

we obtain

$$56\,460 = \frac{(1000)(200) + (TC_e)(50)}{200 + 50}$$

$$56\,460 = \frac{200\,000 + 50TC_e}{250}$$

$$(56\,460)(250) = 200\,000 + 50TC_e$$

$$TC_e = \frac{(56\,460)(250) - 200\,000}{50}$$

$$= 278\,300 \text{ per } 100 \text{ ml (say, } 0.3 \times 10^6 \text{ per } 100 \text{ ml)}$$

So the level of total coliforms permissible in the treated effluent is 0.3×10^6 per 100 ml.

(b) If the raw sewage has a total coliform count of 4×10^8 per 100 ml, the required removal rate is

$$\left[\frac{(4 \times 10^8) - (0.3 \times 10^6)}{4 \times 10^8} \right] \times 100 = 99.925\%$$

(c) For A2 treatment, the total coliform limit is 5000 per 100 ml.
We can use Equation S5

$$TC = TC_o \exp\left[-\frac{k_{tc}x}{U_o} \right]$$

to plot a graph of TC versus x (Figure 56).

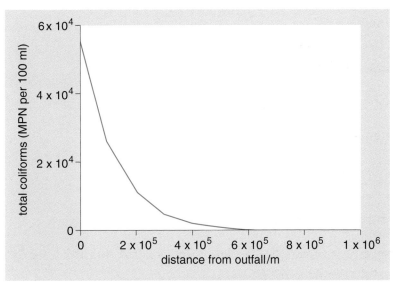

Figure 56

We can then determine the point at which TC is 5000 by concentrating on the region around a TC value of 5000 (Figure 57).

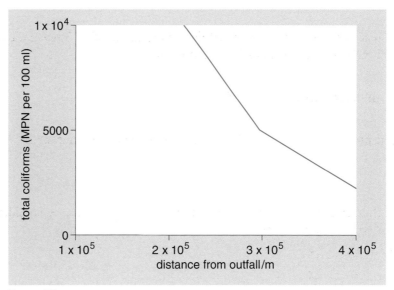

Figure 57

From Figure 57, the location of the water abstraction point is around 300 km downstream of the treated effluent outlet. This is assuming that no other coliform-carrying effluents enter the river after the sewage treatment plant referred to. This is an unrealistic scenario, as there are likely to be not only other sewage treatment works along the river, but also discharges from farms, etc., along the way.

(d) Using Equation S5, with

$TC = 5000$ per 100 ml

$TC_0 = 56\,460$ per 100 ml

$k_{TC} = 0.70$ d^{-1}

$U_0 = 86\,400$ m d^{-1}

$$x = \frac{-U_0}{k_{TC}} \ln\left(\frac{TC}{TC_0}\right)$$

$$= \frac{-86\,400}{0.70} \ln\left(\frac{5000}{56\,460}\right)$$

$$= 299.202 \text{ km}$$

SAQ 62

Pump and intake system to draw water from underground or surface source.

Storage tank to allow sedimentation of gross solids.

Slow sand filter incorporating geotextile fabric.

Floating bowl chlorinator.

Chlorine contact tank.

Distribution lines.

Standpipes.

SAQ 63

A False, since the raw water is pumped through (and therefore flows through at a high rate) and no *Schmutzdecke* is present.

B True.

C False, only compressed air is used.

D True.

E False, at pH 5, undissociated HOCl is at its maximum concentration.

F True.

SAQ 64

(a) Evaporation rate depends on solar radiation, temperature, relative humidity, wind speed, and surface area.

(b) If depth of liquid is d, then

$$1\,\text{m}^3 = 200\,\text{m}^2 \times d$$
$$d = 1/200$$
$$= 0.005\,\text{m}$$
$$= 0.5\,\text{cm}$$

REFERENCES

Binnie, C., Kimber, M. and Smethurst, G. (2002) *Basic Water Treatment*, third edition (first edition 1979), Thomas Telford, London (Set Book).

Davis, J. and Lambert, R. (2002) *Engineering in Emergencies: a practical guide for relief workers*, ITDG Publishing, London.

Franceys, R., Pickford, J. and Reed, R. (1992) *A Guide to the Development of On-Site Sanitation*, World Health Organization, Geneva.

Gardiner, J. and Mance, G. (1984) United Kingdom Water Quality Standards arising from European Community Directives Technical Report TR204, July, WRC.

Harvey, P., Baghri, S. and Reed, R. (2002) *Emergency Sanitation: assessment and programme design*, Water, Engineering and Development Centre (WEDC), Loughborough University.

McJunkin, F. E. (1977) *Hand Pumps for Use in Drinking Water Supplies in Developing Countries*, revised 1983, IRC International Water & Sanitation, Netherlands.

Sundstrom, D. W. and Klei, H. E. (1979) *Wastewater Treatment*, Prentice-Hall, Englewood Cliffs, NJ.

Taylor, W. (1904) *The Examination of Waters and Water Supplies*, J. and A. Churchill Ltd., London.

Tchobanoglous, G. (1979) *Wastewater Engineering: treatment, disposal, re-use*, second edition (first edition 1972), McGraw-Hill, Columbus OH.

United Nations Environment Programme (UNEP)/United Nations Industrial Development Organization (UNIDO) (1991) *Audit and Reduction Manual for Industrial Emissions and Wastes*, UNEP, Paris.

ACKNOWLEDGEMENTS

Grateful acknowledgement is made to the following sources for permission to reproduce material within this product.

Tables

Table 1: Environment and Heritage Service, Department of Environment (2004) 'Standards for the chemical GQA', The Chemical General Quality Assessment Scheme, www.ehsni.gov.uk. Crown copyright material is reproduced under Class Licence Number C01W0000065 with the permission of the Controller of HMSO and the Queen's Printer for Scotland; *Tables 2 & 3:* Environment and Heritage Service, Department of Environment (2004) 'Likely uses and characteristics of classified waters', The Chemical General Quality Assessment Scheme, www.ehsni.gov.uk. Crown copyright material is reproduced under Class Licence Number C01W0000065 with the permission of the Controller of HMSO and the Queen's Printer for Scotland; *Table 4:* Environment and Heritage Service, Department of Environment (2004) 'Standards for the chemical GQA', The Chemical General Quality Assessment Scheme, www.ehsni.gov.uk. Crown copyright material is reproduced under Class Licence Number C01W0000065 with the permission of the Controller of HMSO and the Queen's Printer for Scotland; *Table 5:* 'Phosphate grades and nitrate grades', from General Quality Assessment of Rivers: Nutrients, www.environment-agency.gov.uk. Copyright © The Environment Agency 2004; *Table 6:* 'GQA aesthetics – class tables (two bank site)', from General Quality Assessment of Rivers: Aesthetics, www/environment-agency.gov.uk. Copyright © The Environment Agency 2004; *Table 7:* GQA aesthetics – classification scheme', from General Quality Assessment of Rivers: Aesthetics, www/environment-agency.gov.uk. Copyright © The Environment Agency 2004; *Table 8:* 'Scottish river classification scheme, June 1997', from River Classification Scheme, www.sepa.org.uk. Scottish Environment Protection Agency; *Table 9:* 'River ecosystem classification', from Economic Instruments for Water Pollution, www.defra.gov.uk. Crown copyright material is reproduced under Class Licence Number C01W0000065 with the permission of the Controller of HMSO and the Queen's Printer for Scotland; *Table 18:* 'Typical composition of whey', from Environmental Technology Best Practice Programme, 1988, www/envirowise.gov.uk. Crown copyright material is reproduced under Class Licence Number C01W0000065 with the permission of the Controller of HMSO and the Queen's Printer for Scotland; *Table 19:* 'Annual costs and savings at the plant', from Environmental Technology Best Practice Programme, 1988, www/envirowise.gov.uk.Crown copyright material is reproduced under Class Licence Number C01W0000065 with the permission of the Controller of HMSO and the Queen's Printer for Scotland; *Table 20:* 'Average plant performance data', from Environmental Technology Best Practice Programme, 2000, www/envirowise.gov.uk. Crown copyright material is reproduced under Class Licence Number C01W0000065 with the permission of the Controller of HMSO and the Queen's Printer for Scotland.

Figures

Figure 3: 'Biological removal of iron and manganese', from *Products and Process News*. Courtesy of Degremont, UK Ltd.; *Figure 6:* Hall, T.& Hyde, R.A., (1992), 'The Sirofloc Process', *Water Treatment Processes and Practices*, WRc plc.; *Figure 7:* Metcalf and Eddy Inc., revised by Tchobanoglous, G. (1979) *Wastewater Engineering: Treatment, Disposal, Re-use*, 2nd edn. Copyright 1979, 1992 by McGraw-Hill, Inc. Reproduced with permission of The McGraw-Hill Companies; *Figure 9:* 'The Clivus composting toilet system', from *Technical Information*. Courtesy of Kingsley Clivus Environmental Products Ltd.; *Figure 10:* Franceys, R., *et al.* (1992) 'Operation and maintenance of on-site sanitation', *A Guide to the Development of On-Site Sanitation*, World Health Organization; *Figure 11:* 'The waterless Ecofriendly Biological (WEB) toilet', AC Southwest; *Figure 13:* 'A process flow diagram for a rail maintenance depot'. Reproduced by permission of Nigel Clarke, Enviros Consulting Ltd.; *Figure 14:* 'Turning waste into profit', from Environmental Technology Best Practice Programme, August 1988. Courtesy of Envirowise, www.envirowise.gov.uk. Crown copyright material is reproduced under Class Licence Number C01W0000065 with the permission of the Controller of HMSO and the Queen's Printer for Scotland; *Figure 15:* 'Flow diagram of the effluent treatment plant', from Environmental Technology Best Practice Programme, March 2000. Courtesy of Envirowise, www.envirowise.gov.uk. Crown copyright material is reproduced under Class Licence Number C01W0000065 with the permission of the Controller of HMSO and the Queen's Printer for Scotland; *Figure 16:* 'Water mass balance for the MDF plant (1998 data)', from Environmental Technology Best Practice Programme, March 2000. Courtesy of Envirowise, www/envirowise.gov.uk. Crown copyright material is reproduced under Class Licence Number C01W0000065 with the permission of the Controller of HMSO and the Queen's Printer for Scotland; *Figure 19:* Hatton, E. *et al.*, 'Water quality monitoring using whole cell organisms'. Reproduced by kind permission of WRc Swindon Plc; *Figures 31–34:* Reproduced with permission of Oxfam, GB, from *Water Filtration Pack Instruction Manual*, Oxfam Water Supply Scheme for Emergencies, published @oxfam.org.uk. Oxfam GB, 274 Banbury Road, Oxford, OX2 7DZ; *Figure 35:* McJunkin, F.E. (1977) 'Detail of floating bowl chlorinator', *Hand Pumps: for use in drinking water supplies in developing countries*, Technical Paper Series 10, 1977. IRC International Water and Sanitation Centre; *Figure 36:* McJunkin, F.E. (1977) 'Floating bowl chlorinator, to feed chlorine sollution at a constant rate', *Hand Pumps: for use in drinking water supplies in developing countries*, Technical Paper Series 10, 1977. IRC International Water and Sanitation Centre; *Figure 37:* 'Mobile water purification unit', from www.stella-meta.com. Courtesy of Stella-Meta, part of ITT Industries Inc.; *Figure 38:* 'The filtration mocdule and its cleaning sequence', from www.stella-meta.com. Courtesy of Stella-Meta, part of ITT Industries Inc.; *Figure 40:* Sundstrom, D.W. and Klei, H.E. (1979) 'Relative amounts of HOCl and OCl in water at 20c', *Wastewater Treatment*, Prentice-Hall, Inc.; *Figures 41–43, 45 & 46:* Illustrator: Rod Shaw, WEDC Publications; *Figure 44:* Davis, J. and Lambert, R. (1995) 'A simple grease trap', *Engineering in Emergencies: a Practical Guide for Relief Workers*, I.T.D.G. Publications Ltd.

Every effort has been made to contact copyright holders. If any have been inadvertently overlooked the publishers will be pleased to make the necessary arrangements at the first opportunity.

INDEX

A1 water 20, 166–7
A2 water 20, 166–7
A3 water 20, 166–7
abstractions, control of 40
Accentrifloc clarifier 68
accidental spillages 115
acrylamide/acrylate copolymers 60
action plans 120–1
adsorption 30, 71
aeration 28, 53
aesthetic quality 12–14, 14–16
agriculture, sludges and 124–5
air flotation 68–9
air stripping 30
algal growth 53–4, 145–9
algal toxins 33
aluminium, removal of 28–9
aluminium salts 58, 59
ammonia 8, 17
    removal 33, 37, 93, 94
ammoniacal nitrogen 14, 15
analysis of water 107–22
    remote monitoring of water quality 109–13
    sampling 107–9
    UK emergency procedures 114–22
arsenic 34
artificial water bodies 46, 47
audit monitoring 108–9, 177–8
*Audit and Reduction Manual for Industrial Emissions
    and Wastes* 100
automatic monitoring system 128
average score per taxon (ASPT) 10–11, 14, 15

back-siphonage 118–19
batch chemical processes 92
Bathing Water Directive 42, 149
bentonite 60
biochemical oxygen demand (BOD) 8, 14, 15, 17
    modelling 131–7
biological activated carbon (BAC) 30
biological assessment 9–11, 14, 15
Biological Monitoring Working Party (BMWP)
    score 10
biological removal processes 28–9
biologically active surface mat (*Schmutzdecke*) 73, 154
biosensors 111–13
blue-green algae 33, 53–4, 145–9
brainstorming 97
breaches of regulations 22
breakpoint chlorination 76, 77

Camelford incident 114, 116–17
Camp number 56
canal restoration programme 47
carbonaceous BOD 131–6
catchment control 25
charge neutralisation 55, 58
charging 51
    for trade effluents 82–6, 180–3
check monitoring 108–9, 177–8
cheese-making plant 100–3
chemical assessment 8–9, 14, 15, 23
chemical oxygen demand (COD) 83
chemical parameters 168–9, 174–6
chloramines 75, 76, 77
chlorine/chlorination 19, 32, 53, 75–8, 116
    before filtration 77
chlorine concentration models 130
chlorine contact tank 155–6
chlorine dioxide 32, 75
Churchill–Elmore–Buckingham correlation 140,
    141, 143
clarification 63–70
    settling basins 67–70
    settling velocity 63–7
Clivus system 89–90
closed-loop control 61
coagulation 55–62, 154
    coagulants 58–60
    control of 61
coliform bacteria 118, 149–52
colour 12, 13
    removal of 27
complete water treatment model 131
composting toilets 89–91
computational fluid dynamics (CFD) 129
computer modelling 126–52
    complete water treatment model 131
    river modelling 126–8
    river quality modelling 127–8, 131–52
    underground waters 128–9
    water distribution system 130–1
Consultants in Communicable Disease Control
    (CCDC) 120–1
consultation 46
consumer complaints 22
continuous chemical processes 92
continuous composting toilet 90
Control of Pollution Act 1974 (COPA) 82
copper 17
cost savings 102, 103, 105–6

critical deficit 138–9, 141–5
critical distance 138–9, 141–5
Crown Privilege 83
cryptosporidiosis 31, 121
Cryptosporidium 21, 25–6, 31–3, 117
customer's property 119

Dangerous Substances Directive 39
dechlorination 76, 77
destratification 53
deterministic mass balance 127
developing countries 19, 26
dewatering 123
diatomite filters 74
diffusion 71
direct filtration 69
Directive Concerning the Quality Required of Surface
    Water Intended for the Abstraction of Drinking
    Water 20–1, 39, 107, 108, 166–7
Directive on Pollution Caused by Certain Dangerous
    Substances Discharged into the Aquatic
    Environment 19–20
Directive on the Quality of Water Intended for Human
    Consumption 21, 168–72
    indicator parameters 171–2
    mandatory parameters 168–70
discharge
    to rivers 81, 82
    to sewers 81, 82, 82–6
disinfection 26, 32, 34, 75–80
    kinetics of 77–8
dissolved oxygen 8, 14, 15, 17
    modelling 127–8, 137–45
distribution system
    modelling 130–1
    problems in 117–19
dosing pumps 116
double layer compression 55, 58
drag force 63–4
drainage, natural 162
Drinking Water Directive 21, 25, 107, 108
Drinking Water Inspector (NI) 23
Drinking Water Inspectorate (DWI) 21–3, 31, 121
Dynasand filter 74

ecological quality index (EQI) 10–11, 15
ecological status 41, 47–8
economic assessment 98–9
effluents, trade 81–8
    charging for 82–6, 180–3
    control 81–6
    sampling 86–8
    see also industrial wastewater treatment
electrical power failure 115–16

electrodialysis 35
emergency procedures 114–22
    distribution system 117–19
    pollution of raw water 114–15
    problems with customer's property 119
    statutory notifications and action plans 120–1
    treatment works 115–17
emergency water supply systems 153–65
    emergency sanitation 160–5
    portable water purification equipment 157–60
    possible treatment system 153–7
entrapment in precipitates 55, 58
Environment Agency 7, 43, 44, 115
    river quality monitoring systems 18
environmental assessment 99
Envirowise programme 94
Escherichia coli (*E. coli*) 112, 113, 118
estuaries 128
European Union Directives 19–21, 39
    see also under individual directives
eutrophication 53, 145–9
evaporation pans 164
evapotranspiration beds 164–5

faecal indicators 118
    see also coliform bacteria
ferric salts 58
ferric sulphate 59
fertilisers, sludges as 124–5
Fibrotex expandable bed filter 32
filtration 69, 71–4, 117, 157–8
    membrane technology 26, 35, 101–2, 104–5
fine screens 52
fish monitors 109–10
floating bowl chlorinator 155, 156
floc-blanket clarifiers 69
flocculation 55–62
    control of 61
flood forecasting 126
fluoride, addition of 34
foam 12, 13
full treatment and recycling 93
fuller's earth 60

general aspects monitoring 50
general pollution monitors 109–11
General Quality Assessment (GQA) scheme 8–14
    aesthetic aspects 12–14
    biological assessment 9–11
    chemical assessment 8–9
    nutrient assessment 11–12
Giardia 32–3
granular activated carbon (GAC) 30
grease traps 163

groundwaters 25–6
   classification 48
   modelling 128–9
   monitoring 50
   pollution 114–15
   treatment 36
Guide (G) values 20, 166–7
gulls 53

hardness 17, 34
hazardous substances 49
heavily modified water bodies 46–7
heavy metals 94
history, and water quality 18–21
horizontal-flow sedimentation tanks 67
hypochlorite ion 76
hypochlorous acid 76

Imperative (I) values 20, 166–7
in-stream nutrient concentrations 146–9
Incident Control Team 120
inclined plates 67
indicator parameters 171–2, 176
industrial wastewater treatment 81–106, 179–83
   case studies 100–6
   trade effluent control 81–6
   trade effluent sampling 86–8
   waste minimisation see waste minimisation
initial dissolved oxygen deficit 137–8
interception 71
International Cleaner Production Information
      Clearinghouse (ICPIC) 97
investigative monitoring 49
ion exchange 33–4
ion-selective electrodes 107
iron, removal of 28–9
iron salts 58, 59
irrigation 165

jar tests 58, 60, 61

Langelier Saturation Index 19
lead 26, 33
leakage 19, 23
life-cycle assessments 99
Liljendahl system 88–9
linked computer models 128
linuron 112–13
litter 12, 13

magnetite 61–2
mandatory parameters 168–70
manganese, removal of 28–9
Marine Pollution Monitoring Management Group 45

mass balance 95, 105, 127
medium density fibreboard (MDF) plant 103–6
membrane technology 26, 35, 101–2, 104–5
Merlin monitoring system 18
metering 19
microbiological parameters 168, 173
microfiltration 35
microstrainers 27, 52, 54
mixing 55–8
modelling see computer modelling
Mogden formula 83–4, 181–3
monitoring 84
   automatic monitoring system 128
   check and audit monitoring 108–9, 177–8
   remote monitoring 109–13
   river quality monitoring for pollution 18
   waste minimisation programme 99–100
monitoring programmes 49–50

nanofiltration 35, 101, 102
national standards 19
Natura 2000 list 42, 50
natural drainage 162
network analysis models 130–1
nitrates 11–12, 33–4, 37
nitrogen 124–5, 146–9
   ammoniacal 14, 15
nitrogenous BOD (NBOD) 136–7, 145
Northern Ireland 7, 23, 40, 44, 82
   river classification 16
Northern Ireland Environment and Heritage Service
      7, 44
Northern Ireland Water Service 7, 82
notifications, statutory 22, 120–1
number of determinations 87
nutrients
   assessment 11–12, 15
   modelling 145–9
   plant nutrients in sludges 124–5

ochre 12, 13
O'Connor–Dobbins correlation 140, 141
odour 12, 13, 34
oils 12, 13, 94
on-site electrolytic chlorination (OSEC) systems 79
one-bank sites 12
operational monitoring 49
organic compounds (or organics) 29–31, 41–2, 94
organic matter 124, 125
Organic Pollution Alarm (OPAL) system 111, 112
Outbreak Control Team 120
overflow rate/surface loading rate 66
Owens–Edwards–Gibb correlation 140, 141
Oxfam emergency water treatment system 153–7

oxidation 28
oxygen
    BOD *see* biochemical oxygen demand
    COD 83
    dissolved *see* dissolved oxygen
ozone 27, 32, 75, 79
ozone contact tank 129

parametric values (PVs) 168–72
partial treatment and reuse 93
particle bridging 55, 58
payback period 99
pesticides 29–30
pH 14, 15, 17
phenols 94
phosphates 11–12, 37
phosphorus 124, 125, 146–9
pit latrines 160–2
plant nutrients 124–5
pollution
    definition 51
    raw water 114–15
    river quality monitoring for 18
    see also emergency procedures
pollution pulse, modelling 128
polyacrylamide 59, 60
polyelectrolytes 59–60
polymerised aluminium and iron salts 59
portable water purification equipment 157–60
potassium 124, 125
prescribed concentrations and values 173–5
preservation and storage of samples 107–8
pressure filters 73
pretreatment 52–4
priority substances 41–2, 49
privatisation 19
process flow diagrams 95, 96
process modification 92
protected areas 41–2
    monitoring 49–50
Protection of Groundwater Against Pollution Caused
    by Certain Dangerous Substances Directive 20, 39
public information and consultation 46
Public Health Act 1961 82
Public Health (Control of Disease) Act 1984 120–1
Public Health (Drainage of Trade Premises) Act 1937
    82
Pulsator tank 68

Radioactive Substances Act 1960 83
rainwater run-off 165
rapid gravity sand filters 54, 72, 72–3
reaeration coefficient 140–1
recycling 93

reference conditions 48
regulations 23
remote monitoring 109–13
reuse of water *see* water reuse
reverse osmosis 35, 101, 102
Reynolds number 63–4
risk assessment 44–6
River Basin Management Planning System 43–6, 48
River Ecosystem classification 16, 17
river modelling 126–8
river quality 8–25
    GQA 8–14
    modelling 127–8, 131–52
    monitoring for pollution 18
    Northern Ireland 16
    Scotland 14–16
riverbank wells 153, 154
rivers, discharge to 81, 82
RIVPACS (River Invertebrate Prediction and
    Classification System) 9–10, 14, 16

sampling
    trade effluent 86–8
    water analysis 107–9
sanitation, emergency 160–5
saturation concentration of dissolved oxygen 138
Schmutzdecke 73, 154
Scotland 7, 23, 40, 44, 82
    river classification 14–16
Scottish Environment Protection Agency (SEPA) 7, 44
Scottish Office Department of Agriculture and
    Fisheries, Water Services Unit 23
Scottish Water 7, 82
scum 12, 13
sedimentation 71
settling basins 67–70
    practical considerations and choice 68
    selection of basin type 69–70
    types and operation 67–8
settling velocity 63–7
sewage conveyance vacuum system 88–9
sewage fungus 12, 13
sewage sludge 124–5
    co-disposal of waterworks sludge with 124
    constituents valuable for agriculture 124–5
sewage treatment 37–8, 52
    emergency sanitation 160–2
    on site 89–91
Sewerage (Scotland) Act 1968 82
sewers, discharge to 81, 82, 82–6
Sherlock monitoring system 18
Sirofloc process 27, 32, 61–2
slow sand filters 72, 73, 153, 154–5
sludge-blanket clarifiers 69

sludges 123–5
soakaways 162–3
sodium dichloroisocyanurate 158–9
sodium hypochlorite 79, 157
solids, removal of 26–7, 37, 94
specific substance monitors 111–13
spillages, accidental 115
statutory notifications 22, 120–1
statutory water quality objectives (SWQOs) 16–17
sterilisation 75
'sterilisation' tablets 158–60
Stokes' law 65
storage and preservation of samples 107–8
storage tanks 153, 154
straining 71
stratification 53
streaming current detector (SCD) 61
Streeter–Phelps equation 137
subsurface dams 153, 155
sullage 89, 162–5
superchlorination 76, 77
surface loading rate/overflow rate 66
surface waters
    classification 47–8
    monitoring 49
    pollution 114–15
    quality characteristics of surface water intended for
        the abstraction of drinking water 20–1, 166–7
    river quality see river quality
Surface Waters (Abstraction for Drinking Water)
    (Classification) Regulations 1996 21
surveillance monitoring 49
Synechococcus 112, 113

taste 34
taxa 10–11
Technical Annexes 41–3
technical assessment 98
technical audits 21
terminal velocity 64
tertiary sewage treatment 37
thickening 123
three-day reports 22
timber products plant 103–6
toilets 89–91, 160–2
toxic substances 15
'Trade Effluent Discharged to the Sewer:
    Recommended Guidelines for Control and
    Charging' 82–4, 179–83
    charging for trade effluents 180–3
    Consent Conditions 180
    Directions 180
    objectives of trade effluent control 179
    operation of scheme of control 179

trade effluents see effluents, trade
treatment works, emergencies at 115–17
    see also water treatment
trihalomethanes (THMs) 30–1, 75, 77
tube clarifiers 67
turbidity 26, 69
two-bank sites 12, 13

ultrafiltration 35, 101, 102
ultraviolet radiation (UV) 75, 79
underground waters see groundwaters
uniformity coefficient 72
upward-flow settling basins 68
Urban Waste Water Treatment Directive 53

vacuum system of sewage conveyance 88–9
velocity gradient (*G* value) 56–7
ventilated improved pit latrine (VIP) 161–2
viscosity 63
volatile organic compounds (VOCs) 30
Vyredox process 28

waste audits 95–7
    manuals for 100
    methodology 95–7
waste minimisation 88–106
    case studies on water reuse and 100–6
    reducing wastage of water in industry 91–3
    treatment options to enable reuse 93–4
waste minimisation programme 95–100
    continuous monitoring and targeting 99–100
    dissemination of results 100
    evaluation 98–9
    implementation of waste minimisation measures 99
    option generation 97–8
    review audits 100
    waste audit 95–7
wastes, waterworks 123–5
Water Act 1973 82
water-borne illness 120–1
water by-laws 23
water demand 23
Water Framework Directive 19, 21, 39–51
    environmental objectives 46–7
    general aspects 51
    groundwater classification 48
    key factors 40–6
    monitoring programmes 49–50
    surface water classification 47–8
Water Resources Act 1991 16
water reuse 92–3
    case studies of waste minimisation and 100–6
    treatment options to enable 93–4
water 'sterilisation' tablets 158–60

water storage reservoirs 52–3
Water Supply (Water Quality) Regulations 2000 21,
    173–6
water treatment 25–38
    complete water treatment model 131
    emergencies at treatment works 115–17
    emergency system 153–7
    general principles 26–34
    groundwaters 36
    membrane processes 35
    similarities and differences between sewage
        treatment and 37–8
waterless electronic biological composting toilet 90–1
Watersure on-line pollution monitor 111
waterworks wastes and sludges 123–5
wedge wire screen 52
weighters 60
wells, riverbank 153, 154
whey recovery system 101–3
wholesomeness 18–19
worker participation 98
World Health Organization (WHO) 26
    Guidelines for Drinking Water Quality 19
wound-fibre filters 32, 74
WRc Fish Monitor 109–10

zinc 17

T308 COURSE TEAM

Stephen Burnley   *course team chair/author Wastes/Modelling*

Ernie Taylor   *course manager*

Rod Barratt   *author, Air*

Sylvan Bentley   *picture researcher*

Sophia Braybrooke   *editor*

Philippa Broadbent   *buyer, materials procurement*

Rozy Carleton   *course secretary*

David Cooke   *author, Wastes/Modelling*

Daphne Cross   *assistant buyer, materials procurement*

Jonathan Davies   *graphic designer*

Tony Duggan   *learning projects manager*

Jim Frederickson   *critical reader, Wastes*

Toni Gladding   *author, Wastes*

Keith Horton   *author, Project Guide; critical reader, Noise*

Karen Lemmon   *compositor*

James McLannahan   *critical reader, general, Water*

Lara Mynors   *media project manager*

Suresh Nesaratnam   *author, Water*

John Newbury   *critical reader*

Stewart Nixon   *software designer*

Janice Robertson   *editor*

David Sharp   *critical reader, Noise*

Lynn Short   *software designer*

Shahram Taherzadeh   *author, Noise*

Howie Twiner   *graphic artist*

James Warren   *author/critical reader, Air*

In addition the course team wishes to thank the following for reviewing the material:

External assessor   Professor E. Stentiford, University of Leeds

Block 1   Brian Buckley, consultant

MAIN TEXTS OF T308

An Introduction to Modelling

Block 1   Potable Water Treatment

Block 2   Managing Air Quality

Block 3   Assessing Noise in Our Environment

Block 4   Solid Wastes Management